Managing Mathematical Projects
– with Success!

Springer
London
Berlin
Heidelberg
New York
Hong Kong
Milan
Paris
Tokyo

P.P.G. Dyke

Managing Mathematical Projects – with Success!

With 59 Figures

Springer

Philip P.G. Dyke, BSc, PhD
Professor of Applied Mathematics, University of Plymouth, Drake Circus, Plymouth, Devon, PL4 8AA, UK

Whilst we have made considerable efforts to contact all holders of copyright material contained in this book, we have failed to locate some of them. Should holders wish to contact the Publisher, we will be happy to come to some arrangement with them.

The cartoons in Chapters 1 and 2 and on the front cover are by Noel Ford © 2002. www.fordcartoon.com

British Library Cataloguing in Publication Data
Dyke, Philip
   Managing mathematical projects – with success!
   1. Mathematics – Study and teaching (Higher) 2. Mathematics –
Study and teaching (Higher) – Case studies 3. Project method
in teaching 4. Project method in teaching – Case studies
   I. Title
   510.7'11
ISBN 1852337362

Library of Congress Cataloging-in-Publication Data
Dyke, P.P.G.
   Managing mathematical projects – with success! / P.P.G. Dyke.
     p. cm.
   Includes index.
   ISBN 1-85233-736-2 (alk. paper)
    1. Mathematics – Study and teaching (Higher) 2. Project method in teaching. 3.
Mathematics – Study and teaching (Higher) – Case studies. 4. Project method in
teaching – Case studies. I. Title. II. Series.
   QA20.P73D95 2004
   510'.71'1—dc22   2003060692

Apart from any fair dealing for the purposes of research or private study, or criticism or review, as permitted under the Copyright, Designs and Patents Act 1988, this publication may only be reproduced, stored or transmitted, in any form or by any means, with the prior permission in writing of the publishers, or in the case of reprographic reproduction in accordance with the terms of licences issued by the Copyright Licensing Agency. Enquiries concerning reproduction outside those terms should be sent to the publishers.

ISBN 1-85233-736-2 Springer-Verlag London Berlin Heidelberg
Springer-Verlag is a part of Springer Science+Business Media
springeronline.com

© Springer-Verlag London Limited 2004
Printed in the United States of America

The use of registered names, trademarks, etc. in this publication does not imply, even in the absence of a specific statement, that such names are exempt from the relevant laws and regulations and therefore free for general use.

The publisher makes no representation, express or implied, with regard to the accuracy of the information contained in this book and cannot accept any legal responsibility or liability for any errors or omissions that may be made.

Typesetting: Camera-ready by author
12/3830-543210 Printed on acid-free paper SPIN 10867616

To Heather

Preface

This book is intended for both undergraduate students and their staff as a guide to individual projects, group projects and case studies in mathematics. It covers all aspects of setting up projects and their assessment. The bulk of the text is devoted to giving worked examples of the various kinds of project. The author has benefited from a dual career first in an engineering department where he spent eight years which included looking after individual and group projects, often of a very practical nature. For the last twenty years, he has been within mathematics but still supervising different kinds of projects. This book is therefore written largely from experience. Mathematics here excludes statistics and operational research.

In the last few years there has been pressure on all undergraduate courses to include student centred learning. In mathematics this is difficult because most of the curriculum is devoted to learning mathematical skills. These skills are invariably assessed by examination as it is only in the controlled environment of an examination that the assessors can be sure who is being assessed. One does not assess skills such as spelling or piano grades by setting coursework or projects, hence many would ask how can you assess mathematics other than by a closed book examination? My reasoning is that with the increase in access the nature of many mathematics degrees has changed. If mathematics as a discipline is to keep pace with other undergraduate subjects it has to include modules that enable the student to develop in the broad sense. This means students must have the opportunity to work by themselves on a mathematical topic, or the opportunity to work with three or four other students towards a common goal. It is recognised that this must not be compulsory as there must be room for the clever student who has mathematical flair and may go on to a Fields Medal. This is the mathematics equivalent to the Nobel Prize. Incidentally there is a story that there is only no Nobel Prize in mathematics because

Alfred Nobel's wife eloped with a mathematician! As Nobel never married, the story cannot be true. However, perhaps he lost out in love to a mathematician. It is also said that Nobel had an intense personal dislike of the Swedish mathematician Gösta Mittag-Leffler who was attractive to the ladies and would have been the obvious candidate for the Nobel prize in Mathematics. Whatever the truth, it is a sad omission but as it happens many who have graduated in mathematics have gained a Nobel Prize–Paul Dirac (Physics 1933), Bertrand Russell (Literature 1950) and Richard Feynman (Physics 1965) to name but three.

These days mathematics must not be seen as an elitist subject only to be taken by specialists, but as a useful subject which many can benefit from studying. The usefulness of mathematics should be obvious to scientists, engineers and economists particularly as even the softer areas of these subjects become more quantitative, but its service as a basic discipline that aids the underlying thought processes needs emphasising too. Mathematics graduates have gone on to be politicians and lawyers as well as to succeed in the creative arts. The cry of "what use is Pure Mathematics" should be ignored; who in their right mind questions the use of music, literature or fine art? Mathematics has always been the most transferable of skills, but mathematicians have not traditionally been exposed to project work and group work as undergraduates. Traditionally, the mathematics undergraduate has had no forum in which they can discuss approaches to a problem with fellow students, or to voice their own opinions. This is no longer acceptable, and this text will help both students and lecturers see how projects and case studies can work well in mathematics.

In the text, there are passages that are verbatim extracts from student projects. These are indented and in a smaller font. In other places, the text may state that the following "was done in the project" or some such phrase. In these instances, the author has not copied the student's work but has used the same mathematical method or proof so that the reader understands how the student has approached a particular piece of mathematics. In all cases the author would like to thank all the students who have contributed projects and group work to this text, particularly those who he has supervised. Their anonymity has been preserved although they will easily recognise their own projects one hopes. Three individual projects have been included in their entirety in three separate appendices. Heartfelt thanks go to the three students concerned for permitting this, especially those for whom the project experience was less than ideal. The prime purpose of this text is to learn from the examples of others and to encourage students to try projects and case studies. One can often learn much from reading the works of other students, so these examples are very valuable. For the record, these projects have been faithfully reproduced as closely as possible from the originals. This means that the figures were scanned in and therefore deliberately do not match the usual high standards of Springer-Verlag. On the other hand, the use of LaTeX has beautified the look of the mathematics, if not its content!

Thanks also go to the staff of the Mathematics and Statistics Department at the University of Plymouth. Many of the ideas, particularly those on assessment come from them. I thank those publishers who have permitted me to use figures from their books. Finally a big thank you to Noel Ford who drew all the cartoons to my sometimes less than well formed ideas.

Phil Dyke
September 2003

Contents

1. **Introduction and Organisation** 1
   1.1 Individual Projects 3
   1.2 Group Projects 7
   1.3 Case Studies 10

2. **Assessment** 11
   2.1 Introduction 11
   2.2 Interim Reports 12
   2.3 Verbal Presentations 13
   2.4 Final Report 21
   2.5 Moderating 23
   2.6 Assessment of Case Studies 24

3. **Individual Projects** 27
   3.1 Introduction 27
   3.2 Selecting a Project 29
   3.3 Report Writing 31
   3.4 Non-Euclidean Geometry 34
     3.4.1 Scope 35
     3.4.2 Project Details 36
   3.5 Boomerangs 44
     3.5.1 Scope 45
     3.5.2 Project Details 47
   3.6 Hurricane Dynamics 52
     3.6.1 Scope 53
     3.6.2 Project Details 55

	3.7	Hypergeometric Functions	59
		3.7.1 Scope	60
		3.7.2 Project Details	61
	3.8	Summary	65
	3.9	Project Examples	66
4.	**Group Projects**		**73**
	4.1	Introduction	73
	4.2	Setting up Group Projects	74
		4.2.1 Peer Assessment	74
		4.2.2 Dividing into Groups	76
	4.3	Estuarial Diffusion	77
	4.4	Graphs and Networks	82
	4.5	Fourier Transforms	89
	4.6	Orbital Motion	92
	4.7	Conclusion	95
	4.8	Further Suggestions	96
5.	**Case Studies**		**105**
	5.1	Introduction	105
	5.2	Ocean Surface Dynamics	106
	5.3	Non-linear Oscillations	116
	5.4	Traffic Flow	120
	5.5	Contour Integral Solutions to ODEs	128
	5.6	Optimisation	134
	5.7	Euler and Series	141
	5.8	Summary	146
	5.9	Exercises	147
A.	**Project Example 1: Topics in Galois Theory**		**151**
	A.1	Galois' Approach	153
		A.1.1 Preparation	153
		A.1.2 The Galois Resolvent	157
		A.1.3 The Galois Group	161
		A.1.4 Soluble Equations and Soluble Groups	165
	A.2	The Modern Approach	169
		A.2.1 Field Extension	169
		A.2.2 The Galois Group	173
		A.2.3 Applying Galois Theory	177
	A.3	Soluble Groups	180
		A.3.1 Normal Subgroup Series	180
		A.3.2 Normal Subgroups	181

- A.3.3 Simple Groups 183
- A.3.4 p-Groups 183
- A.4 Geometrical Constructions 184
  - A.4.1 Constructible Points 185
  - A.4.2 Impossibility Proofs 187
  - A.4.3 Performing Algebraic Operations by Construction 188
  - A.4.4 Regular n-gons 188

B. Project Example 2: Algebraic Curves 195
- B.1 Basic Definitions and Properties 197
  - B.1.1 Complex Algebraic Curves and Real Algebraic Curves .. 197
  - B.1.2 Projective Spaces 198
  - B.1.3 Affine and Projective Curves 200
  - B.1.4 Singular Points 200
- B.2 Intersection of Two Curves and Points of Inflection 202
  - B.2.1 Bezout's Theorem 202
  - B.2.2 Points of Inflection on a Curve 207
- B.3 Conics and Cubics 208
  - B.3.1 Conics 208
  - B.3.2 Cubics 208
  - B.3.3 Additive Group Structure on a Cubic 212
- B.4 Complex Analysis 214
  - B.4.1 Holomorphic Functions and Entire Functions 214
  - B.4.2 Closed Curve Theorem and Line Integrals 215
  - B.4.3 Liouville's Theorem and Fundamental Theorem of Algebra 216
  - B.4.4 Properties of Holomorphic Functions 217
  - B.4.5 General Cauchy Closed Curve Theorem 219
  - B.4.6 Isolated Singularities and Removable Singularities . 219
  - B.4.7 Laurent Expansions 221
  - B.4.8 Residue Theorem 221
  - B.4.9 Conformal Mapping 224
- B.5 Topology and Riemann Surfaces 225
  - B.5.1 Topology of Complex Algebraic Curves 225
  - B.5.2 Riemann Surfaces 227
  - B.5.3 Degeneration of a Cubic 231
  - B.5.4 Singularities and Riemann Surfaces 233
- B.6 Further Topics 234
  - B.6.1 The Weierstrass Function 235
  - B.6.2 Differential Forms on a Riemann Surface 236
  - B.6.3 Abel's Theorem 238

C. Project Example 3: Water Waves on a Sloping Beach 241
- C.1 Abstract 241
- C.2 Introduction 242
- C.3 Surface Waves 242
  - C.3.1 The Current 243
  - C.3.2 The Boundary Conditions 244
  - C.3.3 A Separable Solution of Laplace's Equation 246
- C.4 [No Title] 248
  - C.4.1 The Velocity of the Waves 248
  - C.4.2 The Group Velocity of the Waves 248
  - C.4.3 The Motion of the Particles 248
  - C.4.4 Breaking Waves in Shallow Water 249
- C.5 [No Title] 252
  - C.5.1 Plane Waves 252
  - C.5.2 Wave Rays 253
  - C.5.3 The Waves Approaching a Beach 255
  - C.5.4 Wave rays in shallow water 257
- C.6 [No Title] 259
  - C.6.1 Conclusion and Discussion 259

Index 263

1
Introduction and Organisation

It is only in the last twenty years or so that mathematics courses at degree level have been assessed by anything other than examinations. At school level in the UK, GCSE Mathematics has always contained continuous assessment as part of the portfolio since the introduction in the late 1960s of its predecessor the CSE. This was controversial at the time and remains so amongst a significant number of teachers and academics, especially those in higher education in mathematics. The main problem lies in the accuracy of the assessment. Cheating is easy and difficult to detect unless the assessments are carefully designed. Much of mathematics is a skill acquired by practice much like playing a musical instrument or participating in a sport. No-one would assess a concert performance using continuous assessment. Of course other more discursive subjects seem to have always had projects and case studies. Mathematics itself can be partitioned into various sub-categories: of these only statistics, operational research and to some extent numerical methods have taken naturally to this kind of assessment. In fact, excluding only the most theoretical probability, it can be argued that the whole *raison d'etre* behind statistics is to get meaning from data, to analyse, infer and predict, and data sets come from a wide variety of sources such as biology, economics, sport, etc. Operational research too is all about finding optimal routes, the best strategies, algorithms for the most compact packing and so on and so requires input of data from subjects that are outside mathematics. Therefore projects and case studies are natural vehicles for assessing the ability of the student or group of students to apply statistical or operational research techniques. The rest of mathematics is rather different, so much so in fact that projects and case studies centred around statistics and operational research have been explicitly excluded from this text. Applied mathematics has a different problem. If the modelling effort shifts to be focussed entirely on the problem, the discipline risks becoming labelled as engineering, applied physics or the subject in whatever pigeon hole the problem lies. Given a real application, applied mathematics can cross the border. In fact there are national differences and most of what is termed physical applied mathematics in the United Kingdom lies outside mathematics in the rest of Europe and the USA. In the USA, differential equations are applied mathematics, but mechanics, fluid mechanics and electromagnetic theory are parts of physics. Here we retain the broader UK definition of Applied Mathematics. Projects in pure mathematics are still very much a rarity, but are becoming more commonplace. It is not impossible to base projects around the subjects of algebra, analysis, number theory, topology, geometry or differential equations, with the latter two being an extremely rich source. For example algebra with number theory is the basis of codes and cryptography. The first two appendices give examples of projects based on pure mathematics.

An external reason for moving away from examinations towards other forms of assessment lies in the changing nature and pattern of work. Talking directly to students, it seems more and more likely you can now expect to change jobs much more frequently than your equivalents ten or twenty years ago. It is therefore the duty of University courses to educate you in such a way that you can cope with this. If all the assessment was in the form of examinations, the tendency would be (in mathematics) for you to train yourself to be able to do a range of problems and regurgitate these to answer examination questions. There would be no need to communicate with others, not even fellow mathematicians. No need to be able to describe in layman's terms a piece of mathematics. No need to write precisely and understandably. Yet it is these primarily non-mathematical skills that are required in interview situations and in situations where you need to convince others of your worth. There are now many more graduates, with the number edging towards 50% of the population in the UK. Therefore it is essential that all students have as part of their degree something that prepares them for the real world, and for that all important job interview. Not just the interview of course. Once successfully past the interview stage, these qualities also form important employment skills. Of course, a good degree is one mark of distinction, but an unusual project or case study is icing on the cake as it gives you something extra to enthuse about at an interview or in discussions with prospective employers.

A detailed discussion of assessment is covered in the next chapter. There are various types of projects and case studies which are now briefly described in terms of their organisation. It is expected that with experience and with the changing nature of higher education and associated technology, other forms of project and case study will emerge that combine some of the features outlined below as well as incorporate new ones.

1.1 Individual Projects

The essence of an individual project is that it is a piece of work completed by just one person. The advantages of this are obvious. The topic can be geared to the desires of an individual, special interests can be catered for, you can work at your own pace and on completion there is a project report that you can be proud of that is all your own work. There are, sadly, some disadvantages too. Overworked academics can be hard to find for advice, especially given the larger student/staff ratios, all too prevalent these days in the continual drive for greater efficiency. Individual projects can be a lonely enterprise; it is difficult to share triumphs and disasters with fellow students, there is no opportunity for co-operation as there might be in tutorial exercises or other coursework. However even with these disadvantages the individual project is now a common and popular part of the undergraduate mathematics curriculum.

The individual project is a firm part of most final year degrees in other subjects. Some would say that it is arguably the most important part. In engineering, the degree would not be accredited without it for it gives the opportunity for the student to show understanding and flair as well acting as the focus for many of the topics covered in the formal lectures. It has assumed particular importance these days as modular courses are now common, and these can lead to an over-compartmentalised curriculum. The project then becomes a vital glue for the final year modules. In vocational subjects it can also relate the curriculum more closely to the world of work and help the student to get that all important first job. These statements are also true for mathematics. There is no doubt that many a final year project has helped to steer a student into a particular career. In a degree that has a sandwich year, which is a year spent in industry usually just prior to the final year, the final year individual project can be successfully used to integrate this experience into the degree. A successful sandwich year can also lead to that all important first job.

Individual projects in any year other than final is rare, and in my experience unknown in mathematics. They are not considered here.

Sometime towards the end of the second year of a full time mathematics degree that contains an individual project, a handout will normally be issued to all students. This can be in the form of a list of available staff and their interests, or it can simply be a list of available projects. My personal preference is to give the student some flexibility, but also some guidelines in terms of general

Figure 1.1 Overworked academics can be hard to find

topics available. A project too well defined begins to resemble just another lecture course: on the other hand a project designed completely by the student is invariably over-ambitious. In Plymouth I would be rich if I had £10 every time a student has approached me with a desire to do "something to do with sailing or windsurfing"! The important point is for you to discuss potential projects with one or more potential project supervisors. Ideally you will have chosen the project quickly enough to be able to do preliminary work over the summer break, at least theoretically! However, this is not essential and may be impractical as many of you are far away indeed from relevant reference material, computers or even a pencil in the summer months. One final but important point about year long projects: some staff need to be reminded that the project is undergraduate and that it lasts for less than a year and is (usually) less than 20% of the curriculum. It is not postgraduate and it is definitely not an MSc or a PhD! Your personal tutor is a good choice to issue this reminder: if your project supervisor and personal tutor are one and the same try another sympathetic ear in the Department, going to the Head of Department if necessary. Assessing individual projects forms part of the next chapter, and examples of projects are given in Chapter 3.

Sometimes individual projects last less than a whole academic year. In fact it is not rare for a project to last one semester which is half an academic year. In this case, the work has to be telescoped into half the usual duration. The above advice still applies, with the comment that it is important to remember the reduced time available. The subject of the project must be chosen promptly, and help and guidance with regard to the management of the project needs to happen within the first few weeks of commencement. Normally, supervisor and student meet weekly to monitor progress, and it is always important to make these contacts purposeful, no more so than for a project that only lasts half a year. Whatever, it is important that students have a quiet place to work which is free from distractions when writing their project. If your digs or student room is noisy, explore the possibility of booking somewhere in the library, or it might be possible to get a small room somewhere on campus set aside for project students.

Some courses offer "mini-projects". These will be short pieces of work that last two or three weeks maximum. Each student is set a task to do within this short time span and at the end submits a report of five to eight pages. This only differs from standard coursework because each student is set or chooses something different. The report is the only assessment, and it is usually worth no more than 20% of the module. This kind of mini-project is not explored here.

In the examples of Chapter 3, attention will be restricted to the year long and half year projects.

Figure 1.2 Students should work in an environment which is free from distractions

1.2 Group Projects

Group projects are perhaps less common, and are certainly very rare in mathematics. There are many reasons for this, the most apparent being the difficulty in assessing individual contributions to group projects with accuracy. The whole question of assessment will be addressed in the next chapter.

Nevertheless group projects do exist and for at least one good reason. They provide an excellent vehicle for co-operative work. Mathematics is not renowned for encouraging work in groups: almost uniquely mathematicians can be very successful working by themselves in total isolation. Arguably the best and most innovative work is still done this way. A recent example is the proof of Fermat's Last Theorem by Andrew Wiles. However ordinary mortals who are the bulk of students reading mathematics will normally end up working alongside other professionals (who will not be mathematicians but perhaps engineers, scientists, business graduates or graduates in one of the humanities) in commerce or industry. The group project thus provides a valuable introduction to working and getting along with others. Interesting questions are how big to make the group, and how to choose who is in each group. Groups of three, four or five are probably ideal. I have also worked with a group of two and a group of six. The group of two was quite successful, but the scope of the project was limited: many hands make light work! On the other hand a large group can result in the quieter members being overlooked and partially squeezed out of the final group project report. This could result in the student being unsatisfied with the experience: too many cooks spoil the broth perhaps! Some lecturers allow groups to form naturally. This has the merits of giving the student the opportunity (and responsibility!) for the make up of the group but can lead to an uneven spread of ability. All the good students in one group is less of a problem than having a group of struggling students. On the other hand, forcing students into a group could be counter-productive and merely lead to disputes and non-participation. My own preference is to allow the students to organise themselves into groups for group projects. It is important in group projects for the tutor to play a very active role at the beginning. Students would usually be unfamiliar with the details of a group project, so a handout such as the one shown would be a good idea.

The square brackets after each assessment item would give the actual date these were due. The module leader would distribute this handout to the whole class. This would be followed by an introduction to each group project topic with perhaps a forty minute run through outlining the type of mathematics that might be contained in each project together with some ideas on the possible tasks each member of the group may undertake. Some discussion would then take place about which were likely to run, whether there were any project ideas

Figure 1.3   Forcing students into a group could... lead to disputes

Group Project

Organisation

A set of students, minimum number 2, maximum number 6, preferred number 3 - 5, work as a team on a project. This is called a Group Project.

At the beginning of the module, there will be a short sequence of lectures explaining each project. At the end of the lectures, students will be expected to choose one of them to work on. This does not preclude students from coming up with their own ideas for projects, but any ideas must be vetted for suitability by the module leader, and there must be a group of students willing to work together on it. In practice each project consists of a number of topics that individuals work on. There is a handout accompanying each project that includes suggested topics, however these are only suggestions and students working on the project are at liberty to formulate their own strategies for completing the project. Again, the module leader is there to help and provide guidance on, for example equalising workload within a project group. The goal is to produce a project report by the end of the period to which all the group has contributed more or less equally, and with which all members of the group is proud to be associated.

The module is worth 10 credits (there are 120 credits in the year) so a rough guide is that each student should spend 100 hours on their contribution.

After the initial lectures, formal class contact is restricted to one hour per week when the module leader is available in classroom for general consultancy, making general points from matters that arise, giving advice, etc. There are two other timetabled hours in the week when the module leader is available in his/her office. This should be used for specific queries, usually but not necessarily of a technical nature.

The choice of who does a particular project is largely up to you. You will be asked to state preferences, and these will form the basis of the split into groups. Note that it is perfectly possible for more than one group to work on the same project. Usually the projects have enough scope for many groups. The module leader does have the final say as it may be in your own interest to ensure that the composition of each group has roughly equal mathematical ability.

Assessment

The assessment of this module will be as follows:

1. Contribution to final report [hand in deadline]

   70%

2. Interim Report [hand in deadline]

   10%

3. Verbal report to the whole class [date of orals]

   10%

4. Peer assessment by other members of the group.

   10%

from the student body, and some further discussion on peer assessment. This latter topic may the subject of a further handout (see the next chapter). All this is to give the students enough information to enable them to make up their minds. Some specific examples of group projects are given in Chapter 4.

1.3 Case Studies

The case study varies in concept and execution. One possibility is that the case studies module of a mathematics degree consists of formal lectures and assessments much like any traditional module on, say, differential equations. Each case study is perhaps spread over two or three lectures and students are assessed on their understanding of it either then and there with a test or coursework, or at the end of the module via an examination. This kind of case study module is not all that different from a standard lecture course. There is a kind of case study module that is distinctive which might run as follows. A series of lectures is given at the beginning of the academic period (term or semester). These are a mixture of the technical and the managerial. Technical lectures introduce each case study whilst the management lectures tell the students about how they will be assessed and how they will interact with the rest of the group. Typically each case study is very different: one may have arisen from an industrial problem whereas another may be research paper based, and a third can be a mathematical topic which is not new to mathematicians, but is new to the students and different from the rest of their course. Some may even involve a group of students working off campus with confidential data! This however is rare in mathematics.

The case study can be run in a similar way to a group project, but here there is a stronger case for choosing groups for the first case study then rotating the membership for a new one. For both case studies and group projects, there may be student concern over equality of workload. For example those in a group whose project concerns analytical dynamics or quantum mechanics may feel that their project is harder than one on graphs and networks. My view is that of course all projects and case studies are different and therefore some are harder than others; however the assessment is adjusted accordingly. The potentially serious problem is where there is one very strong student in a group of less able students. This student could end up doing most of the work on the one hand, or not achieving his or her potential on the other. Some of these difficulties are faced when we consider assessment in the next chapter.

2
Assessment

2.1 Introduction

Assessment is the most important aspect of education to get right at all times, but especially when running a project or case study. These days students are not backward in telling you when things are going wrong, and if it involves their marks so much the worse. It is also true that now it seems more difficult to get students to do anything out of altruism than it used to be. It was once the case that students would expect "homework" to be set regularly, be marked and returned regularly. This was seen to be a standard part of any lecture course in mathematics. Nowadays, if homework is set too many students will not do it, even if this is accompanied by an apology and a Harrison Ford like lopsided grin. Despite it being helpful, a fact often duly recognised by the students themselves, the fact that it is not *essential* means that it does not get done. Or more accurately homework tends not to get done by those who would benefit most by doing it. Although educational theorists still talk about formative and summative assessment, in practice the distinction between the two is often fuzzy. After the first few weeks, everything students do usually needs to be connected to a mark that contributes to the final grade. Or, at the very least, be seen to be extremely useful towards gaining marks. In this chapter we tackle the all important question of how to assess individual projects, group projects and case studies, although it is the first two that take the lion's share of the chapter.

An important aspect is to make sure the student is committed quickly after the start of the module. One way of ensuring this is to set an assessment quite near the beginning. Such an assessment need not be particularly summative, i.e. it need not contribute much toward the final grade, just enough to convince recalcitrant students to do something. It is largely formative and is there to help the student to know he is on the right lines. We call this assessment the "interim report".

2.2 Interim Reports

Interim reports are found in group projects and in individual projects. Generally they are absent from case studies. The group project, as the name implies, is a group of students working towards the production of a common piece of work. On the other hand the individual project is done by one person, with greater or lesser assistance from the supervisor. The role of the interim report is therefore subtly different in each case so let us discuss them separately.

To begin with individual project interim reports, these are primarily an indication that progress is being made with the project. Advice on how to write them is postponed until the next chapter. Here we shall concentrate solely on assessment. The supervisor should not be surprised by its contents, but it provides the student with an opportunity to review progress. It should also be assessed by another member of staff who can provide an independent view. The contents can be technical, outlining the mathematics done so far and signalling future work. Or it can be a general overview, describing what has been done. It is largely formative, but it is a good idea for it to attract some mark which is worth up to 10% of the total project mark. This mark should be the average given by the two assessors.

Group project interim reports are a little different. They are individual efforts, perhaps the only part of the group project that falls into this category (but see the next section on verbal presentations). Once more, guidance on how to write an interim group project report is found in Chapter 4, although it is worth looking at the guidance in Chapter 3 too as there are many similarities. The purpose of the interim report is still to assess progress so far, but it is also to help each student to identify his or her role within the group. A group project is based on the collective work of a group of students. In any particular group there may be dominance. One student may be physically dominant through appearance, a booming voice, a domineering attitude, etc. More likely one student could be intellectually superior, the boffin of the class perhaps. Another possibility is that the group consists of a "clique" plus an outsider. It is all too easy for this outsider to feel ostracised and leave the project for the rest to do.

In all of these cases, circumstances could lead to one or more members of the group not contributing sufficiently. The interim report forces every member of a group to home in on a sub-topic, do some work on it and write up what they have done. It also gives the module leader an indication of what each member of a group has done and will do, hence providing a valuable oversight of the entire project. Each interim report should still be second marked, although the final mark in this case could be agreed (over coffee perhaps) rather than a straight arithmetic mean. The mark given to the interim report here could be 15% rather than 10% to emphasise its importance. It is very important in both cases to give positive feedback to each student. At this early stage it is all too easy for students to become disheartened about the project or to go off at a tangent, or to get stuck. The feedback from the interim report is a useful vehicle to help the student formally. More seriously, if there is a severe problem later (for example health related), evidence from this stage can be very helpful.

2.3 Verbal Presentations

These days it is important for every student to get the opportunity to give a presentation. The era of the bright graduate in mathematics or engineering with a first class degree who can only mumble incoherently in interviews or meetings has passed. Being good on paper and in passing examinations is now only part of the story. All forms of project and case study can provide good vehicles for students to get used to presenting to others. Let us run through different ways this can be done and outline assessment procedures.

The most obvious, and the most daunting for the student, is to prepare a talk for the whole class. This is most often done in the context of an individual project where the student is given the opportunity either half-way through the year or at submission time to tell everyone about it. It has the advantage of being something that is new to the bulk of the audience (fellow students) and the staff present are normally sympathetic and do not ask too many awkward questions at the end. Nevertheless it is an ordeal for most students, particularly if they have not done anything like it before. The secret is to prepare well and run through the material a number of times before the event. Use of visual aids is encouraged. These used to be overhead projector transparencies written on or containing photocopied writing and equations, but increasingly now include PowerPoint presentations or computer projections and video clips. Here is a short list that students will find useful.

Figure 2.1 It is very important to give positive feedback to each student.

Figure 2.2 Staff are normally sympathetic and do not ask awkward questions.

In a good formal presentation the speaker should:

1. Await formal instructions from the Chairman or Announcer.
2. Thank the Chairman or Announcer.
3. Give a proper formal introduction to the talk in the form "I am .... from .... and my talk is on ....".
4. Always address the audience.
5. Keep the audience interested.
6. Give an outline of the talk at the start by going through a list of contents.
7. Present material in a logical, structured manner.
8. Use appropriate visual aids, usually overhead projector transparencies or PowerPoint.
9. Make sure the information on visual aids is clear and readable, and make sure they are not overcrowded.
10. Keep within the allotted time.
11. Conclude the talk properly (e.g. run through a list of conclusions).
12. Thank the Chairman and audience for their attention.
13. Respond clearly, concisely, correctly and politely to questions.

The presentation skills are of course important, but what must be right is the technical content. Students should remember that lecturers can see through the most glamorous presentation and soon detect any flaws or misunderstandings in the material of the project. The assessment of such project talks is done against two principal criteria, the mathematical content and technical level of the talk, and presentation skills. The weighting is either 50% each or 60%: 40% in favour of the content. There may be a heavier weighting in favour of content, but the presentation side should not be completely ignored. Again, these days verbal presentations are double marked to ensure fairness and quality. The verbal presentation is usually about 10% of the total individual project mark. Normally, students find the experience of standing and presenting very nerve wracking, and it is quite useful to be given some general guidelines on how performance relates to marks. The following list matches performance to grade:

2. Assessment

Figure 2.3   Students should keep within the allotted time.

Classification	Presentation
First	Clear, insightful, good answers to probing questions.
Upper Second	Clear explanations, able to answer questions.
Lower Second	Some good explanations and answers, but sometimes unclear or unconvincing.
Third	Acceptable explanations of part of project, partial answers to some questions, but otherwise confused.
Fail	Completely muddled, or missed the point, or did not turn up!

Students do not need convincing of the usefulness of giving a verbal presentation, both from the point of view of learning the material of their project and also in instilling confidence in situations such as the job interview. Many interviews these days include an element of presentation and students who have done this kind of thing before start with a great advantage.

In group projects, there is normally one group project presentation, but all members of the group must participate. This can be all group members taking a turn at the OHP, but it can be more adventurous. For example, a simulated interview, or one student taking on the role of Master of Ceremonies and introducing the rest in turn. This can present difficulties if the amount of time each member of the group is at the podium, so to speak, is very different. The assessor should take account only of the contribution in terms of technical content and presentation and this ought to be independent of the actual number of minutes it takes. The only time the student is penalised is if the presentation is far too long due to lack of discipline or organisation of material, or far too short due to ill-preparedness. Whatever format is used and however long each contribution takes, it is possible to give each participant an individual mark although inevitably there is less differentiation between content marks than between presentation marks.

There are other forms of verbal presentation that are well worth trying. One is the viva-voce, or viva for short. This is best suited to the individual project, but could be tried elsewhere. Normally the student is questioned for about fifteen minutes. The questioning is done by the assessor who may or may not be the project supervisor. In fact, an arrangement that works well is if the questions are led by an independent assessor with the project supervisor taking the role of the informed assessor, mostly listening but chipping in the odd question. It is important that the supervisor is involved in assessing as a final year student often does not do justice to his or her knowledge in the stressful situation of a viva. The presence of the supervisor normally acts to calm the student, although of course one can always think of exceptions! This contrasts with the PhD viva in which the supervisor is definitely "prisoner's friend" and normally takes no role as assessor. Some postgraduate research students elect not to have their supervisor present: this has not been my experience for undergraduate student projects. It is important that there are two assessors for the project, and in the unlikely event that the student objects to one of these being his supervisor, two others have to be appointed. There is no point arguing with the student about this; however it must be made clear that there are severe disadvantages not allowing the supervisor to assess the project. For example the two people assessing the project may be unfamiliar with some of the circumstances of its production and not give credit at the correct weighting for work done. The assessment in this case is unlikely to be accurate. On the other hand to force a student to include an assessor/supervisor who, for whatever reason, is biased is to invite an appeal. Appeals are messy and best avoided. At the close of the viva, the student leaves the room and the two assessors agree a mark. This mark could be for the whole project and not just the student's performance at the viva. Indeed it is important that each student knows the criteria under which the viva is marked. Usually it is presentation skills and depth of knowledge shown. It sometimes happens that a student breaks down under vicious and over penetrative questioning. This is of course most unfortunate and may be the fault of an over zealous assessor who usually bears no malice but gets caught up in the subject matter of the project to such an extent that (s)he forgets that it is only an undergraduate student on the receiving end of questions and not an eminent professor. The supervisor should spot when this is happening and help the student, but if this does not happen there could once again be grounds for appeal. It is useful for a passive third assessor to be present in case of disputes. Finally of course telephone calls, knocks on the door and other external interference need to be eliminated as with any confidential meeting. I once conducted an oral examination with a raging storm outside which, with only partial success, we all tried to ignore!

Normally the viva is not marked separately and if it is, it is only given up to 10% of the marks. Oral presentations on the other hand are usually marked separately, again usually attracting 10% of the marks as a maximum.

Figure 2.4 The presence of the supervisor normally acts to calm the student.

2.4 Final Report

The actual writing of the report for the individual project is a subject tackled in the next chapter. Here we concentrate on its assessment.

Quite naturally, the final report is the ultimate goal and provides the main criterion by which to judge the success or otherwise of the entire venture. If more than one student is involved in producing the project report, then it is usually invidious to try and mark each contribution separately. Far better to give a single mark for the whole written report and use the other assessments to differentiate between students.

Some attention is now given to developing the criteria needed to arrive at a mark. To read a large piece of written work and then to decide "this is worth $x\%$" is really not satisfactory and can leave academic staff open to student appeal. One way to be more helpful is to tell students how to get high marks in the written project. Here is a typical list:

Classification	Comment
First	Elegant, excellent English, well structured, very few typographical errors.
Upper Second	Concise, good English, clear structure, not too many typographical errors.
Lower Second	Acceptable English, some structure, not very well proof-read, rambling prose.
Third	Some acceptable writing, some evidence of planning, many errors, muddled.
Fail	Illiterate and confused.

It is emphasised that these are guidelines and that, for example, not all third class projects will be muddled. There will be some guidance also about the amount of time that should be spent on the project. Ten hours for every mark is consistent with 120 credit points every academic year. Since 1200 hours equates with 400 hours each term, and given a 10 week term, this implies a 40 hour week which is about right. More specific indications of attitude are given in the following list:

Classification	Typical commitment
First	Enthusiastic, worked largely independently of supervisor, found own sources.
Upper Second	Found some extra references, diligent and well read.
Lower Second	Read the given texts thoroughly.
Third	Read the given texts, but was uncritical and sluggish.
Fail	Did not find or read anything relevant.

So far nothing has been said about mathematical accuracy. A very well structured report, well written, thoroughly referenced but full of mathematical errors will not get good marks! Obviously it is also true that projects vary as to their mathematical sophistication. A student who is struggling with the technical aspects of mathematics is unlikely to do well in an individual project such as the hypergeometric functions project outlined in Section 3.7. However a project which is less mathematically demanding may well provide a good vehicle. The thorny question about levels of difficulty then arises. Is it permissible for a student to undertake a project with, in the extreme, no mathematical content? Perhaps the biography of a mathematician? The consensus seems to be that a certain level of mathematics is essential. For example, mathematical biographies, in addition to being scholarly need to contain a reasonably detailed account of the mathematical breakthroughs made by the subject of the biography. If the biography is of an ancient mathematician, say Archimedes or Apollonius then the report needs to contain a substantial amount of the mathematics of finding the area of a parabola using the method of indivisibles, or the geometry of spirals or whatever. An essay on the sociology of the time and some lamentation over all the lost Greek primary sources will not do. Then there are the projects that emerge from the general area termed Mathematics Education. Judging the quality of a project based around the impact of calculators on 'A' level mathematics against that on hypergeometric functions remains very difficult and the subject of some heated debates. The best advice to students is to do some advanced mathematics if at all possible or risk a poor project mark.

Copying has always been a problem to some extent, but in these days of the internet and floppy disc, it is so much easier to do and harder to detect. There are cases of completely plagiarised PhD theses that have gone undetected for years, and I am sure this is also true at undergraduate level. Most universities have in place codes of conduct and severe disciplinary rules which prevent blatant plagiarism. However, the line between research and copying is a fine one especially at undergraduate level. Students must reveal all sources and include them in the reference list, even if this reference is a past project by an ex-student from a few years back. Building on the work of a previous undergraduate is commendable: copying from a previous project is unforgivable.

Although the emphasis above has been on assessing the written individual project report, much of it applies equally well to assessing any project report. The only additional aspect of a group report is assessing the integration of the different contributions. If this is well done, and an external examiner finds it hard to detect where one contribution ends and the next starts then credit should certainly be given. If it is entirely obvious that the report has been written by, say four individuals, then I am not convinced that any penalty should be exacted. On the other hand if each part is done in different type size and font, or there is discontinuity in page numbers (or no page numbers at all), or say two Figure 8's this detracts from the quality of the report and the mark must reflect this.

2.5 Moderating

Moderating is the name given to checking that all marks are fair and reasonable. In projects, this is done in several ways. In an individual project, it is a good idea for there to be a second marker for the project. Both attend any seminar or oral presentation involving the student, and both read and mark the project. The two assessors then agree on a mark, and this becomes the mark submitted to the examination board. In the event of disputes, some institutions involve a third assessor who acts as a referee. Each assessor writes a report on the project, and justifies the mark awarded. If the two grades are widely different, these reports should say why. A third assessor then brings judgement to bear and either agrees with one or other of the assessors, or strikes a happy medium. The third assessor's ruling should be final. If both assessors make similar comments, but award different marks, for example: "superb piece of work" 50% and "superb piece of work" 70% then the criteria given in the section above should be used to remind the first of these assessors that a "superb piece of work" equates to first class honours and ought to be awarded 70% rather than 50%. It is also common for one person, usually the final year tutor or equivalent to read all the individual project reports and check the comparative fairness of the marks. After this has been done, there is often a short(!) meeting at which minor adjustments can be made before marks are finally submitted into the administrative system. Most academic staff are willing to concede that marking projects cannot be done to within an accuracy of three percent, and swapping the order of two projects between say 68% and 71% on the strength of the recommendation of the final year project coordinator's re-reading of both projects is normally not controversial.

Group projects are moderated by another member of staff overseeing all the assessment procedures. The moderator is involved in attending oral presentations and in reading the interim and final group reports. There is no involvement in peer assessment other than being aware (and commenting on if necessary) the procedure used.

2.6 Assessment of Case Studies

Some universities and colleges use case studies to mean group projects, in which case the above means of assessing group projects applies. Here a case study is defined as the presentation of an example of the use of mathematics to solve a specific problem, or a new application of mathematics. Normally it will span three to five hours of lectures. Students are thus attending lectures and tutorials in much the same way as in a standard module. At first sight therefore, there seems little reason to assess case studies in a manner different from a standard lecture course, that is, simply set an examination at the end and support this with some coursework, the weighting being perhaps 80% on the examination and 20% on the coursework. It is however possible to be more adventurous. Since a case study is an in-depth look at a particular mathematical technique, or some (to the student) new branch of mathematics it might be possible to set extended coursework with students working in pairs. Assessment can then involve some elements similar to the group project: verbal presentation and a common report. There should still be an examination as this is the fairest way of assessing the understanding of mathematics that has been given to the whole class. For case studies, an open book examination is particularly suitable, although in my personal experience there should be some restriction on the volume of material allowed into the examination room.

Figure 2.5 There should be some limit on the volume of material allowed into an open book examination.

3
Individual Projects

3.1 Introduction

This chapter is unashamedly directed at you the student. It is designed to help you through tackling the individual project that often forms part of the final year of a mathematics degree. Here some general aspects of undertaking an individual project are introduced and several examples of projects are discussed. The project examples will encompass a fair spread of mathematics, but that is not really the point. This is not a chapter on introducing new mathematics; instead the mathematics of each project example is used to amplify aspects of the process you go through in order to write a successful project. The form the chapter takes is as follows. There are introductory and general sections on starting up and writing a project, then there are the project examples. These examples are taken from real projects that ran in the University of Plymouth, and there is extensive quoting from each example project. Alongside this, there is a commentary on aspects such as relevant advice, technical matters, and assessment.

For many students the hardest part of a project is not the mathematics, although this of course should be challenging and at the correct level. No, it is learning how to write a report that is coherent, well structured, uses good English as well as correct mathematics that provides the challenge. Mathematics students are not renowned for their expertise in writing: for some this reputation is harsh but nevertheless there remains a core of students who pursued mathematics because of their inability to string words together meaningfully.

These days it is not acceptable for graduates to lack transferable skills, and writing is certainly one of these. What better way to get a recalcitrant mathematics student to write than to base a project around a favourite mathematical topic? There is more on these writing skills in Section 3.3. Aside from the actual English there is also readability and structure. It is important to give every student the opportunity to do exciting mathematics of course, but features such as telling a story, giving some biographical detail (where appropriate), spelling out limitations, etc., are equally important. In writing up a project you are putting mathematics in some context and presenting the whole in a coherent package for the intelligent layman. It is emphasised that all the examples here are from real projects that formed part of the final year of a degree course in mathematics. In fact they all formed one sixth of the assessment of the final year. It is useful for you to know and the author is pleased to report that they all succeeded. As in studies of pollution or the spread of disease, it is not acceptable to engineer a deliberate project failure in order to have an example of one! The isolated examples of project failures this particular academic has encountered have all been due to the student opting out. Often this can easily be prevented, so it is particularly sad but thankfully rare when it happens. This is why it is so important to engage the student early and have in place various checks that monitor progress and that enable the lecturer to spot the first signs of things going wrong. Who can blame a student from "opting out" of a project when the subject matter, although glamorous, turns out to be totally beyond their mathematical ability, or who is left totally isolated with a supervisor who is never there? The projects featured here were graded as follows: one lower second, one upper second and two firsts. Details on the assessment of each project follow each description. Three of the examples are taken from the efforts of my own project students. This enables me to be more forthcoming about these assessments.

Mention needs to be made of the role of computing in a mathematics project. A common view is that computing should be regarded as a tool. It is a very powerful tool and facilitates numerical analysis. It is quite legitimate for a student to spend a lot of project time writing computer programs as long as these are a means to an end and not an end in themselves. For example, a FORTRAN program that solves a boundary value problem numerically is a legitimate part of a project. The student (and project supervisor!) must guard against spending all the time honing the program, making it super-efficient or providing glamorous output at the expense of correct modelling. This is particularly so for the less mathematically gifted student who is working on a project without demanding mathematics. One example might be web based material for solving differential equations as an educational tool, undertaken by a student who is pursuing a joint degree of Mathematics with Education. The evaluation of the output from such a project must be in terms of its educational value as a tool for learning mathematics and not in terms of the computing design itself. Of course, the downloading of unreferenced material from the internet is as unacceptable in a project as anywhere else. The experienced lecturer can almost always detect where this is happening (changes of style and the sudden onset of excellence!) Some mention of the difficulties of assessing Mathematics Education projects was made in the last chapter. Of course good design is related to a good educational product but the emphasis must be on the latter in a mathematics degree.

If you are following a joint degree in mathematics and computing say, then a project that combines both disciplines is a good idea. The danger with all joint degrees is that the student is serving two masters with different agendas. It is particularly important to write down in the form of a short description or using bullet points what you are expected to do in this case. In the examples that follow, this is always done. There is no computing oriented project example included here. One is pure mathematics, two are applied mathematics and one is mathematical methods but containing a great deal of analysis. In no way is it possible to cover all the possible designs of projects. Other omissions include projects on mathematical physics, mathematics education and mathematical statistics. However it is not the topic that is the emphasis but the method of doing the project, its order and level. Before starting the project, how do you decide on a topic?

3.2 Selecting a Project

How do you select a project? What are the pros and cons? The student who knocks on the door of the lecturer and says "I want to do a project" needs to be given advice. First, is the project compulsory? If it is not, you will find yourself being questioned as to how you came to the conclusion that the project is better than the alternatives on offer. Having decided on doing the project, do you know what subject? A common first step is to provide all prospective students with a list of staff and their interests. Such a list is given here as Table 1. Of course, you can come up with your own project topic, but this has to be approved by a member of academic staff. Some universities do not allow this and have a prescribed list of projects from which to choose. In one sense this is a good idea as the university is taking all the responsibility and if the project falls over for any reason, the university is in a better position to put things right. The student who is permitted to embark on a rather ill-thought-out project on their pet topic for example "the mathematics of sailing" can be

Table 3.1 Staff Interests

Name	Areas of Interest
Professor Sir Kit Breaker	Electrodynamics
Professor Cuthbert Calculus	Rockets and Astrophysics
Dr. Derek Delta	Quantum Mechanics, Generalised Functions
Professor Phil Dyke	Dynamical Oceanography and Meteorology, Mechanics, Special Functions
Dr. Eugene Function	Boundary Value Problems
Professor Polly Hedron	Geometry
Dr. Esther Mator	Mathematical Statistics
Dr. Archie Meadies	History of Mathematics
Dr. Dai Namics	Mechanics
Professor Ternut Onanov	Boolean Algebra, Discrete Mathematics
Professor Carol Orry	Real and Complex Analysis
Professor Sam Poole	Applied Statistics
Dr. Victor Product	Mathematical Methods
Dr. Algy Rithem	Numerical Methods
Dr. Constance See	Relativity, Mathematical Physics
Dr. Ena Sher	Mechanics
Dr. Ron Skion	Differential Equations
Dr. Fay Splane	Dynamic Systems
Dr. Cary Steristics	Partial Differential Equations
Professor Harry Stottle	Foundations of Mathematics, Philosophy
Dr. Con Straint	Optimisation
Dr. May Tricks	Linear Algebra, Systems of Equations
Professor Eddy Viscosity	Fluid Dynamics, Turbulence

left all at sea!

It is important that the scope is defined and specific tasks are assigned for the student to complete. Personally, I always interrogate the student armed with their performance to date to find out what kind of mathematics they enjoy and what kind of mathematics they are good at. If these are the same, then no problems. The projects supervised by me usually involve my own interests in oceanography and meteorology. This is useful for the prospective project student as a topic that is the research interest of the lecturer provides the student with a plethora of background material, good advice and enthusiasm. I was once in the position of "bailing out" a member of staff and supervising a student on a topic in which I had very little interest; it was not a success and I would not recommend it. The list of names provided in Table 1 is not (with one notable exception) real of course, but their interests are quite typical. This mythical mathematics department has several members of staff interested in aspects of differential equations which probably reflects a research strength; on the other hand nobody professes an interest in graphs and networks or the applied side of operational research. The student must be prepared for this. It is now becoming the norm for departments to specialise, and small departments in particular will not be able to cover the whole of mathematics at the depth required for a final year project. Departments have their strengths and weaknesses and if your particular interest is not catered for you must make a choice between good supervision of a less than ideal topic or a more hands-off supervision of a favoured topic. All the academic staff in the department can do is make sure that you are aware of the situation and to proffer advice. Another choice is whether to go for a project subject that is highly mathematical (say something on quark confinement in modern quantum mechanics) which might be interesting but difficult, or something less mathematical (say a biography of Leonhard Euler). Often the advice is that the less mathematical project is safer, that is it is likely to be graded in the middle (lower second). Therefore if you are striving for a first, go for a hard project that might get a first class mark and help you achieve your ambition. On the other hand, if you are struggling to keep up or finding the technicalities of several of the modules taxing, then go for the mathematically less demanding project at which you are less likely to founder. I think that this is good advice. With greater access to higher education a priority with many governments, notably here in the UK, it is important for mathematics degrees to cater for a wider ability range than they did twenty or thirty years ago.

Before looking at some examples of projects, the next section is devoted to the art of report writing, including the interim report

3.3 Report Writing

Report writing is an art everyone would do well to master. There certainly will be guidelines on how to write a report in the student handbook which is usually given to each one of you during induction week. If this is not the case there will be guidelines issued either on paper to each student or (less satisfactorily given student hardship and the cost of printing) on the university or departmental website. In this section the opportunity will be taken to give some general guidance.

It is common for many lecturers to forget that the project report will probably be the first lengthy scientifically based piece of work a student has ever produced. This being so, although some of what follows may seem obvious to the readers of this text who are academics, in my experience students will find it useful.

In this chapter we concentrate mainly on the production of the individual project report, but most of what follows is general enough to apply to all substantial pieces of student writing. The project usually gets going in early autumn (say October) and has to be completed by early May. There will be departures from this timetable which happens to be the one that is used at the University of Plymouth, but the departures will not be significant except in one case. In some universities an individual project will only extend over half an academic year, either between October and January, or February and June. If this is the case, the general principles remain the same except that all has to be done over a shorter time scale. The supervisor and student should meet weekly. In recent years this meeting has often had to be short, say half an hour, due to resource constraints; however it is the frequency that is much more important than its length. At an early meeting, the project supervisor should give advice on writing the report and map out the writing timetable for the student. You also need to come to grips with the "interim report" if it exists. The temptation to spend all contact discussing the technicalities of the project should be resisted. In conversations with their tutors and lecturers and in written guidelines, students will be told to make good use of their supervisor. Early on, you will be faced with the first point at which some kind of assessment might be expected. This is usually the interim report. If this is as soon as two months into the project, then it is there to act as a mark at which you can spell out what you expect to do in your project in some detail. If some mathematics is required, perhaps to give the assessors some indication of the mathematical level, then concentrate on solving one specific problem and write up the solution. What emerges is a piece of work that both helps you to structure the project and assess how hard the mathematics is likely to be, and helps the assessors get a better idea of what might be coming. If the interim report is expected later, say after Christmas, then it is really a progress report. In it you should give a short account of the work done so far and what you expect to complete before the hand-in date. This interim report provides a rich source for any oral presentation that might also accompany it. Returning to the final report, if this is produced entirely without input from the supervisor it is never the best and is often a disaster. You should give yourself plenty of time to write the final report. A common error is to use all available time to concentrate on the technical material. It is always tempting to go just that little bit further mathematically but then the unfortunate student is left rushing the write up as the submission deadline suddenly looms. Time is needed for the supervisor to read the entire submission and to give constructive feedback on aspects from minutiae such as sentence construction, to the overall structure and layout of the report. There also needs to be plenty of time for proof-reading and detection and correction of typing and other perhaps more serious errors.

You need to understand what you have written. It is all too easy to succumb to the temptation to copy mathematics, a proof or derivation, without fully understanding it. There will be some kind of oral assessment, so students doing this are adopting a high risk strategy. It should be emphasised to all students that they need to comprehend all the mathematics that they put in their project. Any reproduced derivation or proof must therefore be fully understood as well as fully referenced. It is the view of this author that mathematics projects are best produced in LaTeX. However, most students will already have experience using Microsoft Word from school or college and it is quite adequate for their needs. Moreover, LaTeX can be difficult to master quickly especially if you are not a programmer and are not used to typing that is not WYSIWYG (What You See Is What You Get). The treatment of mathematics by LaTeX in terms of storage, layout and flexibility is superior as it is a specifically designed mathematics typesetting package. (This book is being produced, needless to say, using LaTeX.) An added advantage is that certain computer algebra packages such as MAPLE interface with LaTeX successfully. There is no natural interaction with Microsoft Word. However if another Microsoft product such as Excel is used for tables or graphs then using Word works quite well. You must make your own choice: assessors should be concerned with the mathematics not the typesetting, although a good layout must be a plus point.

Students are well advised to produce a skeleton report with chapters and section headings before writing anything substantial. You should then give this to the supervisor who will provide valuable feedback. It is also a good idea at an early stage to write a section or two and get feedback on writing style as well as content. The overall style of the report should be for the intelligent layman who can understand the mathematics. There are plenty of spell checkers and grammar checkers on word processing software these days, but students need to be aware of their limitations. Some of the most common words are absent from the electronic spell checker and the grammar advice can vary from the bizarre to the plain wrong. Students from the UK also need to be aware that most of these electronic aids to writing are based on American English. Using a dictionary and a thesaurus therefore remains sound advice. The first part of the project to write up in any detail is the part with which the student is most comfortable. This usually is part of chapter two or three, the "guts" of the project. The introduction is the last part to write.

References are important in a project, and it is important to get both the content and the layout right. LaTeX has a good referencing facility (BibTeX) that can help but may not be easy for the undergraduate to master quickly enough. Books should be referred to in the following style:

Dyke P.P.G. (2000) *An Introduction to Laplace Transforms and Fourier Series* Springer-Verlag, London Berlin Heidelberg 250pp.

Papers on the other hand should be referenced as follows:

Dyke P.P.G. (1980) On the Stokes' drift induced by tidal motions in a wide estuary *Est. Coast. Mar. Sci.* **11** pp 17 - 25.

Sometimes there is an overall page restriction introduced. This is there to prevent students producing a project report the length of which would be more typical of a PhD thesis. Such over-long reports are difficult to assess fairly, they take too much student time to produce, often to the detriment of their other work, and they tend to be undisciplined. Forcing the student to produce a succinct project report is a good strategy. Fifty pages for a project worth one sixth of a year is a good guide.

In the next sections, we introduce the first of the specific projects. They are outlined in some detail and comments made. Remember, the purpose is to give examples of how projects run, not to introduce new mathematics. Therefore if the mathematical content of a particular project is obscure, no matter. You will still be able to get information about project methodology (though not as much as those who understand everything). In all the examples given below, all commentary is mine. There is a portion of quoted text, usually from the project introduction, and the mathematics is quoted directly from the student project, but without the misprints. The project details are given in reported speech. So the phrases "the student wrote...." and "in the project, the student gives the following proof...." will be found. This is to make it clear what is actually in the project and to distinguish it from something this author would like to have seen there but was missing. It is hoped that this style does not detract from the object of this chapter which is to be of use to students actually undertaking and writing up individual projects. The first project is undeniably pure mathematics.

3.4 Non-Euclidean Geometry

This first example is a project that is concerned with the geometries developed in the early nineteenth century. The level of mathematics may not be thought of as being particularly high, but much of it will be new to the student and that is the point. It is important that this is so of course as the regurgitation of notes from another module is unacceptable. In large departments the detection of this could be a real problem. To maximise the detection, wide publicity must be given by placing lists of project titles, students names and their supervisors on noticeboards or bulletin boards on the local intranet. The project is a vehicle for the student to explore an avenue of mathematics on his or her own; to read new mathematics, to understand theorems and corollaries to them, all independent of other students. As a bonus, at job interviews it can be a great advantage to be able to wax lyrically about *your* project.

In the 1960s, all teenagers at third form level (aged 14 or 15, now called year 9 or Key Stage 3 in the UK) fought their way through the theorems of Euclid. Those that became undergraduate mathematicians were thus very well versed in the geometry of triangles and circles and, more importantly could prove results using the axioms that originated from the *Elements* of Euclid. The idea of proposing axioms and using them in a logical way to deduce theorems and corollaries was therefore met at this important formative age. However since the demise of the popularity of Euclid and his geometry, such methodology has all but vanished. The first proofs students see if they see any are algebraic and introduced in the lower sixth. (Proofs by induction and contradiction are the usual first examples). Logical thinking thus takes a back seat these days, and *geometrical* thinking features hardly at all. Potentially the student who took on this project therefore had to learn about geometric proofs perhaps for the first time. Actually, being of mature years this student had the advantage of a 1960s schooling but it lay deep in the memory banks.

Even though there does need to be an overall minimum standard that has to be achieved it is always important to be mindful of the background of the student when reading or supervising a project. It needs to be remembered therefore that mathematics that was met at school by the lecturer could be new to the student and hence be a legitimate vehicle for an undergraduate student project. It is time now to look at the mathematics in some detail. Of course the object here is not to explore non-Euclidean geometry as in a textbook, but to see at first hand some of the processes of producing a project using non-Euclidean geometry as a vehicle.

3.4.1 Scope

One of the most important aspects of an individual project to decide at the outset is its scope. The enthusiastic student is often over ambitious and if allowed free rein the rest of his or her course will suffer! One way to prevent this is to limit pagination as already mentioned, but this is sometimes insufficient. The advice and guidance of the tutor or project supervisor is essential at an early stage. This topic, non-Euclidean geometry, is not new thus advice is vital. For a newer topic which is more narrowly defined (the use of prime numbers for encryption for example) guidance is easier as the subject tends to be more closely defined. In older mathematics there has been time for the subject to mature and grow and for links between it and other mathematics to be forged. The students therefore is in need of more guidance. For this project, the following objectives would have been appropriate:

- to understand the axiomatic approach to geometry

- to outline the principles of projective geometry

- to understand the Klein view of geometry.

Armed with this kind of minimal guidance, the student has less excuse for straying too widely. However (s)he is now given some references and told to go away, read and explore. The student must also be encouraged to seek advice quickly if there is anything they do not understand.

Quite typically but not essentially for a final year project, the reference material consists of textbooks, e.g. H.S.M.Coxeter's "Projective Geometry" published in 1987 by Springer-Verlag. This reinforces the view that the material is indeed suitable undergraduate level mathematics; however it is not part of lectures. As we have already said if it is all entirely lecture material then it is unsuitable as a project topic. However, a project that extends a final stage module is perfectly acceptable and is in fact a good idea. Non-Euclidean geometry may be a topic in final stage undergraduate mathematics in a few universities but it is only given a mention in passing at Plymouth, the source of this and all subsequent project examples.

3.4.2 Project Details

Most projects need to set the mathematics involved in some kind of context. Particularly in a pure mathematics project a certain amount of history of mathematics becomes necessary. Here is an excerpt from the introduction of this project which shows the student's attempt at doing this:

> From Babylonian times some four thousand years ago, there is evidence of the study and use of geometry. In those times the geometry used was practical, for working out areas of fields, volumes of grain and so forth. The formulation of the mathematical problems was rhetorical and procedures used without proof. The move from procedure to proof was instigated by the Greeks who used the axiomatic approach to formalise their geometry. Records of this

(circa 300 BC) have been preserved in Euclid's *Elements*, thirteen volumes of Greek mathematical thought. From this time to the beginning of the eighteenth century little changed in the basic ideas of geometric theory. There had been, during that thousand year gap, controversy over Euclid's fifth *parallel* postulate with many mathematicians working to find a proof of this postulate. In the sixteenth century some progress was made in projective geometry prompted by the artistic attempts to represent realistically three dimensional space in two dimensions. But it was not until the eighteenth century that significant changes appeared in geometrical thinking.

The student then goes on to discuss the axioms of Euclid as they were criticised by mathematicians of the eighteenth century, particularly Felix Klein. Of course, the phrase "non-Euclidean geometry" needs to be defined. From this it follows that it is necessary to understand something about axiomatic mathematics. There is therefore an introductory chapter on axiomatic methods. It has already been mentioned that the student is unlikely to be very familiar with this kind of mathematics. Historically the "guts" of non-Euclidean geometry focuses on developments in the early nineteenth century; therefore some history of events and biographical details of Lobachewsky and Gauss might be expected. The work of Felix Klein is crucial, especially his linking of geometry and group theory. This work dating from 1872 may be expected to feature prominently. There is also David Hilbert's extension (correction?) to the Euclid postulates. It is well to contrast this historical material with projects in applied mathematics where setting the context of the project in the real world tends to be more important. This will be seen in later examples in this chapter.

Following the setting of the historical perspective, the student needs to do some mathematics. Sometimes it is necessary to remind students that a mathematics project needs to contain mathematics! Even those principally concerned with history (a biography of Euler or Gauss for example) needs to convey the mathematics of the time. This point was alluded to in Chapter 2, and will emerge again. This project states axioms as follows:

- For every point P and for every point Q, P not equal to Q there exists a unique line l that passes through P and Q.

- For every line l there exist two distinct points incident with l.

Such axioms with their formality of language can seem very strange to many students. As a common aim of a project is to be able to present new mathematics (new to the students that is) to fellow students it is quite appropriate to select a topic for a short presentation to the class. An introduction to this kind of formal mathematics is one possible topic for this talk arising from this project. This presentation is often the first formal hurdle the project student faces. Choosing as a vehicle a topic covered early in the project is thus a good idea.

Once the general scene is set, one possible route for this project to take was to concentrate on, say, projective geometry. One distinctive aspect of projective geometry is the lack of obvious definition of distance. Another is the presence of pure geometric theorems. Students appreciate being able to try out new mathematics in calculations. In projective geometry homogeneous co-ordinates are usefully employed to prove results. These results can be proved using other means open to the student, for example vector algebra, but homogeneous co-ordinates are particularly neat. At this point we shall delve into the specific mathematical details of the project. Homogeneous co-ordinates are defined as follows. If we have plane Cartesian co-ordinates (X, Y) then a typical straight line takes the form of the linear equation

$$aX + bY + c = 0.$$

By putting

$$X = \frac{x}{z}, \text{ and } Y = \frac{y}{z}$$

the line becomes

$$ax + by + cz = 0.$$

The triplet (x, y, z) defines homogeneous co-ordinates. Points for which $z = 0$ are called "points at infinity" and $z = 0$ itself is the "line at infinity". That all conics look like ellipses with the "ellipse" being a parabola if it is tangent to the line at infinity and a hyperbola if it crosses this line is intriguing to students and helps retain interest as well as cement an understanding of this form of non-Euclidean geometry.

To give a flavour of one of the geometric proofs, here is a statement and proof of Desargues' theorem as it might appear in this student's project.

Theorem 3.1

Desargues' Theorem

If two triangles are in perspective, then the intersection of corresponding sides are collinear, see Figure 3.1.

Proof Let U be the intersection of the lines of perspective, and let it have homogeneous co-ordinates (α, β, γ). This means that if the two triangles are ABC and $A'B'C'$ then AA', BB' and CC' all pass through U as shown in Figure 3.1. We can arbitrarily set the points A, B and C to have co-ordinates $(1, 0, 0)$, $(0, 1, 0)$ and $(0, 0, 1)$ respectively. The vertices of the other triangle

3. Individual Projects

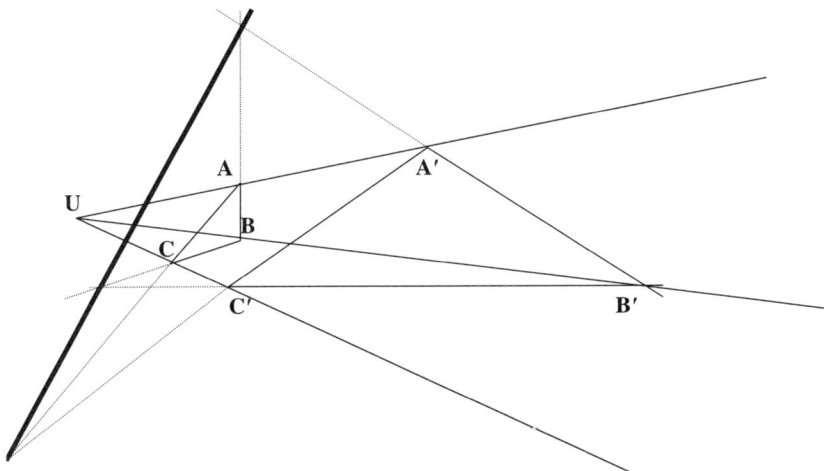

Figure 3.1   Desargues' Theorem.

$A'B'C'$ will be of the form $(\alpha + \lambda, \beta, \gamma)$, $(\alpha, \beta + \mu, \gamma)$ and $(\alpha, \beta, \gamma + \nu)$. The equations of the three sides of $A'B'$, $B'C'$ and $C'A'$ are

$$\begin{vmatrix} x & y & z \\ \alpha + \lambda & \beta & \gamma \\ \alpha & \beta + \mu & \gamma \end{vmatrix} = 0$$

$$\begin{vmatrix} x & y & z \\ \alpha & \beta + \mu & \gamma \\ \alpha & \beta & \gamma + \nu \end{vmatrix} = 0$$

and

$$\begin{vmatrix} x & y & z \\ \alpha & \beta & \gamma + \nu \\ \alpha + \lambda & \beta & \gamma \end{vmatrix} = 0.$$

Algebraic manipulation is used to show that these lines intersect AB ($z = 0$), BC ($x = 0$) and CA ($y = 0$) on the line

$$\frac{x}{\lambda} + \frac{y}{\beta} + \frac{z}{\nu} = 0.$$

Thus we have found in algebraic form the actual line that contain the three points of intersection of AC and $A'C'$, AB and $A'B'$, and BC and $B'C'$. Hence Desargues' theorem is established.

$\square$

This proof demonstrates the power of linear algebra to prove geometric theorems. This proof is devoid of error and its appearance exactly like this in a project might lead one to suspect that it had been copied verbatim from a textbook. This is not a problem provided the student understands what (s)he has copied and this can be assessed via an oral examination. The student must be able to understand for example why it is possible to assign the points A, B and C the rather special looking homogeneous co-ordinates. Any student who was not confident would not be sure this was a correct step and the proof would look "fiddled". The algebraic manipulation used to achieve the final result is not trivial either and a careless student would soon get in a mess. Pure geometry arguments could of course be used, but most students find the algebraic approach easier to follow. These are just a few learning points; I am sure you can spot others.

Let us move on to discuss other technical points. An important quantity in projective geometry is the *cross-ratio* which is defined by the student as follows

$$(A, B, C, D) = \frac{AC}{CB} \div \frac{AD}{DB}.$$

The student points out that only the ratio of lengths is required in this definition and not the lengths themselves. Defining a quantity that is new to the student then examining some of its properties makes a good assignment in a student project. Here is the proof of the invariance of cross-ratio under a $1-1$ correspondence between points as it appears in the project. It is simple algebra but usefully gets the student familiar with the cross-ratio.

Theorem 3.2

Let (x_1, y_1), (x_2, y_2), (x_3, y_3), and (x_4, y_4) be four corresponding pairs of a given $(1, 1)$ correspondence. Then

$$(x_1, x_2, x_3, x_4) = (y_1, y_2, y_3, y_4).$$

Proof Suppose that the equation of the correspondence is

$$axy + bx + cy + d = 0.$$

Then

$$ax_1y_1 + bx_1 + cy_1 + d = 0, \text{ and } ax_3y_3 + bx_3 + cy_3 + d = 0.$$

Subtracting these two equations and rearranging gives

$$b(x_1 - x_3) + c(y_1 - y_3) = -a(x_1 y_1 - x_3 y_3) = -ax_1(y_1 - y_3) - ay_3(x_1 - x_3).$$

This gives
$$\frac{x_1 - x_3}{y_1 - y_3} = -\frac{ax_1 + c}{ay_3 + b}.$$

Similarly we derive
$$\frac{x_1 - x_4}{y_1 - y_4} = -\frac{ax_1 + c}{ay_4 + b},$$
$$\frac{x_2 - x_3}{y_2 - y_3} = -\frac{ax_2 + c}{ay_3 + b}$$

and
$$\frac{x_2 - x_4}{y_2 - y_4} = -\frac{ax_2 + c}{ay_4 + b}.$$

Judicious dividing of these equations to eliminate the constants a, b, and c gives

$$\frac{x_1 - x_3}{x_1 - x_4} \bigg/ \frac{x_2 - x_3}{x_2 - x_4} = \frac{y_1 - y_3}{y_1 - y_4} \bigg/ \frac{y_2 - y_3}{y_2 - y_4}$$

which is the desired result.

$\square$

Of course, pure geometric proofs may also feature. For example here is Pappus' Theorem:

Theorem 3.3 (Pappus)

Let P, Q and R, and L, M and N be two sets of three collinear points. Let QN, RM meet at the point F, let RL, PN meet at the point G and let PM, QL meet at the point H. Then the points F, G and H are collinear.

Proof 1 This is the standard algebraic proof (absent from this project, but given here for information) in outline. Let the six points have the following co-ordinates:

$$P = (p, 0, 1) \quad Q = (q, 0, 1) \quad R = (r, 0, 1)$$
$$L = (l, 1, 0) \quad M = (m, 1, 0) \quad N = (n, 1, 0).$$

The equation of the line QN is the determinant:

$$\begin{vmatrix} x & y & z \\ q & 0 & 1 \\ n & 1 & 0 \end{vmatrix} = 0$$

or
$$-x + ny + qz = 0.$$

The line RM has the equation
$$-x + my + rz = 0$$
which is found similarly. The intersection of these two lines is found by solving the simultaneous equations as
$$F = (mq - nr, q - r, m - n).$$
Similarly
$$G = (nr - lp, r - p, n - l), \text{ and } H = (lp - mq, p - q, l - m).$$
The determinant formed by these right hand sides:
$$\begin{vmatrix} mq - nr & q - r & m - n \\ nr - lp & r - p & n - l \\ lp - mq & p - q & l - m \end{vmatrix}$$
is zero as is apparent by adding the rows. This shows collinearity and establishes Pappus' Theorem.

□

The handle turning nature of this proof meets with favour from the student, but tends to leave the tutor cold. Lecturers prefer geometric proofs that give more insight. Here is one for Pappus' Theorem that resembles the proof that actually appeared in the project report:

Proof 2 In this proof, co-ordinates are not used. Instead we label the intersections of the lines PM and RL as A and PN and RM as B. We then rely on theorems involving projectivity (which may not be familiar to everyone!). A projectivity is a $1-1$ correspondence between points P and Q such that each P gives rise to a unique Q and vice versa. It is sometimes also called a homography. The line $PHAM$ is in projectivity with the line $PQRO$ with respect to the point L. Moving our attention to the point N, $PQRO$ is in turn in projectivity with the line $BFRM$. Therefore the lines $PHAM$ and $BFRM$ are in projectivity, and a theorem can be used that tells us that provided the lines have a point that corresponds to itself (in our case the point M) we have a projectivity and a perspectivity which implies that the lines PB, HF and AR are concurrent. This implies that G lies on HF being the intersection of all three lines and Pappus' theorem is once again established. (See Figure 3.2.)

□

3. Individual Projects   43

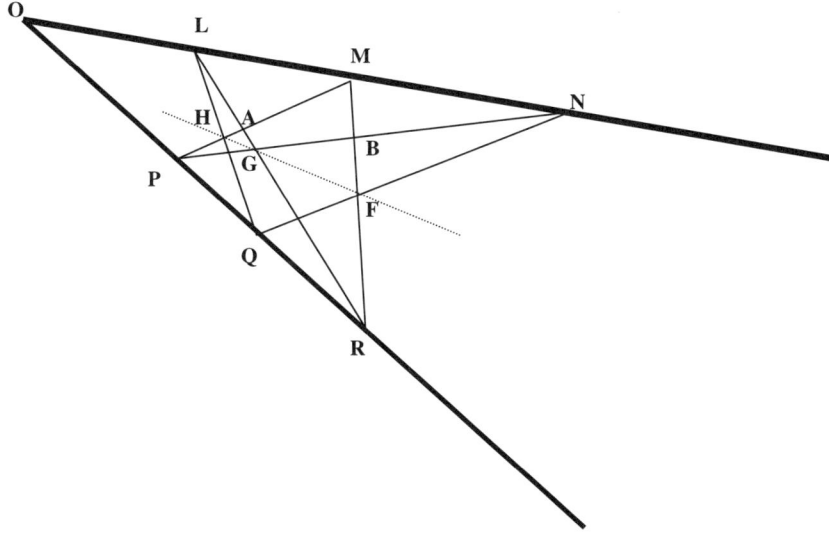

Figure 3.2   Pappus' Theorem.

Which proof is preferred is a matter of taste. The geometric proof demands a specialist background but wins in elegance. The algebraic proof has the advantage of being easier to follow in common to most "methods" proofs. For the student of mathematics, as we have said there is much to recommend the geometric proof and it was probably at the instigation of the project supervisor that this version appears in the project that provided the source material for this section. This kind of geometric thinking is likely to be very new to most students but very beneficial to their mathematical understanding.

There are many other fundamental concepts that might find their way into a project on this topic. For example this student introduced the *Principle of Duality* whereby in any theorem in plane geometry simply swap "point" for "line" and vice versa, for "line joining two points" substitute "intersection of two lines" and vice-versa. One can see for example that the dual of Desargues's theorem is the same as its converse. There are other cases where the application of the Principle of Duality gives rise to distinctly different theorems. There is the whole question of conics which as we have said in projective geometry all look like ellipses, and their relationship with triangles. For example some readers may be familiar with the nine-points circle which in projective geometry becomes the eleven-points conic. Suffice it to say that projective geometry is a very rich subject and many final year projects could be based upon it. If there is a projective geometry module (or two!) somewhere else in the course so much the better as the project can build on the taught material and be much richer as a result. Note that there is not much mention here of the third bullet point under "scope" concerning the Klein view of geometry. As a project progresses it is not unusual for the emphasis to alter and in this case it moved into projective geometry and away from the work of Klein. As long as tutor and student agree this is not a problem. This mirrors the business world in which although strategic plans for the next ten years are drawn up, typically they change inside three.

The assessors for this project were impressed by the mathematics in the project, and by the performance of the student at the presentation and at the oral examination. The layout of the project and the writing style was excellent. It was felt that the student could have done more on the Klein view of geometry, but the consensus final mark was that the project deserved first class honours, but near the borderline.

Some students would rather tackle a project on a subject that relates more to the world outside the classroom. The next project is certainly one of these.

3.5 Boomerangs

The boomerang is the Australian weapon that if thrown correctly moves in a (large) circle, or slightly distorted circle, and lands at the foot of the thrower unless interrupted by hitting the target (a kangaroo perhaps). So if the student wants to find out how a boomerang works, the supervisor needs to tell the student about the kind of mathematical background required. This particular project is very difficult to do within mathematics courses that contain no mechanics, although not impossible. It is all too easy for a specifically defined project such as this ending up as an introduction to rigid body mechanics, particularly as in Plymouth, where this project was set, the course at the time only contained particle mechanics. This provides a less than ideal background. If some knowledge of rigid body dynamics, particularly the spinning body, had been met before then so much of this project need not have been spent on the fundamentals of rigid body mechanics.

The role of applied mathematics in school mathematics has certainly changed a lot in the past thirty years. Mechanics used to be a substantial and compulsory part of the mathematical diet of the 16 to 18 year old. The two thick books by Humphrey and Topping called "Statics" and "Dynamics" and now long out of print covered the material. Old well thumbed copies may still be found in libraries, but more likely they have been sold off cheaply, given away or even thrown out. In any case the material in old 'A' level mechanics texts is almost all inaccessible to today's students due to the lack of geometry already alluded to. It is perfectly possible for the 16 year old to embark on a mathematics course, 'A' level or its equivalent, that contains no physical applied mathematics. Engineers think this sacrilegious, but it is perfectly reasonable if one considers the destiny of students of 'A' level mathematics these days. There is business and finance, and computing that soaks up most students and these need statistics and decision mathematics much more than knowledge of Newton's laws. There are also the social sciences, biology and psychology, all of which have become more quantitative and require knowledge of statistics. Where better to put this statistics than in 'A' level mathematics? Of course there is still a place for traditional applied mathematics, but it is no longer seen as needing to be compulsory. This then has changed the scope of mechanics in mathematics degrees. Some universities can still deliver modules in Mechanics not a great deal different in content to those of the 1960s (apart from computer based material of course). Students from these are well equipped to tackle a project on boomerangs and such universities are large and usually have substantial engineering and/or physics departments where mechanics is also core and provides an environment which is attractive to the academic applied mathematician. When the project started, it soon became apparent that it would be a good idea to incorporate frisbees too. Frisbees are discs which, being symmetric and geometrically simpler than boomerangs are easier to analyse.

3.5.1 Scope

In applied projects it is very rare to feature any history. It is hardly appropriate to give a history of the boomerang. To give a brief historical development of the mechanics and fluid mechanics would be straying too far from the central theme of the project which is not mechanics or fluid mechanics but an application of them. It is of course necessary to have introductory material, but this takes the following different form:

> We tackle the analysis of rotational motion with reference to the frisbee and boomerang. Since the swept circle of the boomerang resembles a frisbee we can base most of our theory on the frisbee then relate the results back to the boomerang. There will obviously be some difference in the motion of the two projectiles and this will be explained in Chapter 6. In what follows we use aerofoil theory to account for the frisbee's ability to hover. For this we

start with irrotational flow past a circular cylinder, using Milne-Thomson's Circle Theorem to obtain the complex potential. We then prove that there is a force $-\rho u \Gamma$ in the y-direction. If we can find a shape for which $\Gamma < 0$ then we can make this force act in the positive y direction. In pursuing this shape we use the Joukowski transform to turn the cylinder into an ellipse, then a flat plate before finally settling on a wing shape with a sharp trailing edge.

So what we have here is not history but a genuine introduction to the subject matter of the project. The language used reflects the inexperience of the student which of course is perfectly acceptable. The material emphasises the fluid mechanics side of the work which most excited this particular student. However given the foregoing remarks about mechanics, this project may also contain an introduction to rigid body mechanics. First of all, a rigid body is defined. The special place of plane rotation then leads to considering rotating bodies which naturally leads to defining the moment of inertia. The description then moves on to defining torque as the moment of force and stating that it is equal to the rate of change of angular momentum, and then progressing to a discussion of three dimensional rotation and gyroscopes. Only those points relevant to spinning boomerangs are covered, but the knowledgeable reader is at once aware of the pace of learning here. For some less mechanically inclined students this would need to be quick and focused. One essential difference between a taught module and a project is that a project is focused and specific whereas a module (on mechanics say) needs to provide reasonable coverage of the subject and thus must be broader. To cover all of mechanics from defining the rigid body to Euler angles and gyroscopes is a full module, but to do that which is necessary for a project on boomerangs can be considerably less. For the project student here, it was about 30% of the contents of a standard mechanics module. The project is principally about modelling a flying aerodynamic body and thus must contain some fluid dynamics. In particular enough background needs to be covered to understand lift. If the student knows complex variable theory then the Joukowski transformation becomes an application of (known) two dimensional inviscid flow theory. Otherwise, some fluid dynamics of an elementary nature needs to appear in the project. As indicated in the excerpt from the introduction, the choice of this particular student is not to include elementary two dimensional flow in the project. In fact there is barely enough for the reader (and writer!) to understand the complex potential, and this is a criticism. Reading the above it is obvious that the student here did know a little fluid mechanics and complex variable theory, but some revision was required. Here then are the aims of the project as agreed between supervisor and student:

- to outline the theory of rigid body mechanics in enough detail to understand how a flying boomerang (or frisbee) remains stable in flight

- to understand aerodynamic lift and apply the theory to boomerangs (or frisbees).

- to understand why the thrown boomerang returns to the thrower.

At the outset, the student was pointed to texts on mechanics and fluid mechanics that covered the topics mentioned above. More specialist articles on frisbees and/or boomerangs were found using library searches. For the talk the student gave an overview of the project as a whole with some details about the Joukowski aerofoil. The amount of mathematics presented was not large, but was enough for members of the audience to get the gist. The talk was very successful. A success almost guaranteed by this particular student's flair for presentation.

3.5.2 Project Details

In this section it is emphasised that the mathematics is quoted directly from the project. As is quite typical this project gets straight on with the mathematical detail, so let us do the same. From the definition of angular momentum ($\mathbf{h} = \mathbf{r} \times m\mathbf{v}$) the project student takes this definition and generalises it to n particles

$$\mathbf{h} = \sum_{i=1}^{n} \mathbf{r}_i \times m_i \mathbf{v}_i$$

where r_i, m_i and v_i are the position vector, mass and velocity of the ith particle respectively. For a rigid body that is rotating with angular velocity $\boldsymbol{\omega}$ we have $\mathbf{v}_i = \boldsymbol{\omega} \times \mathbf{r}_i$. Inserting this into the expression for angular momentum gives

$$\mathbf{h} = \sum_{i=1}^{n} \mathbf{r}_i \times m_i (\boldsymbol{\omega} \times \mathbf{r}_i).$$

The right hand side of this expression contains a triple vector product which can be expressed as follows:

$$\mathbf{r}_i \times (\boldsymbol{\omega} \times \mathbf{r}_i) = \boldsymbol{\omega} r_i^2 - \mathbf{r}_i(\boldsymbol{\omega} . \mathbf{r}_i) = \boldsymbol{\omega} r_i^2,$$

as $\boldsymbol{\omega}$ and $\mathbf{r}_i$ are perpendicular. Thus

$$\mathbf{h} = \sum_{i=1}^{n} \boldsymbol{\omega} m_i r_i^2 = \boldsymbol{\omega} \sum_{i=1}^{n} m_i r_i^2 = I\boldsymbol{\omega},$$

where I is the moment of inertia of the rigid body. The usual steps whereby the finite number of particles become infinitely many infinitesimal particles was skipped for this project. As we have said, this is acceptable; after all the emphasis is on describing boomerangs here. A project report is different from a set of lecture notes as they tell a story and the story can presume, state without proof, etc., leaving out what is not essential to the narrative at the author's discretion.

For boomerangs (and frisbees) the relationship *torque = rate of change of angular momentum* is important. Using the above, this is

$$\mathbf{r} \times \mathbf{F} = I\dot{\boldsymbol{\omega}},$$

where $\mathbf{F}$ is the applied force. To this the student needed to add the analysis of a spinning object. The route to this is via Euler's equations for a body moving freely in three dimensions. Again, this may be part of a lecture course, but less commonly so these days. Using the above notation Euler's equations are quoted in the project in the form:

$$\dot{\mathbf{h}} + \boldsymbol{\omega} \times \mathbf{h} = \mathbf{r} \times \mathbf{F}.$$

Quoting this is fine in one respect, but the student has to understand that equations need to be referred to a *fixed* origin and *fixed* axes. To derive such equations is not trivial and an understanding of Coriolis acceleration and rotating systems is really an essential requirement. Once this is established, the precession of a flying spinning object can be analysed. One route to this is to analyse a fixed spinning object first (a top or gyroscope say, in this project the student used a top) through the following mathematics. Axes that rotate with the body are defined by $(\mathbf{i}, \mathbf{j}, \mathbf{k})$ where $\mathbf{k}$ is the axis of rotation and the other two axes ($\mathbf{i}$ and $\mathbf{j}$) rotate with the top. Figure 3.3 shows the details. Now, the bold Greek letter $\boldsymbol{\omega}$ is reserved for the rotation of the system referred to fixed axes yet to be defined, so the student uses $\boldsymbol{\Omega}$ to denote the angular velocity of the top about its axis of symmetry. (S)he supposes further that this axis is at a constant angle α to a fixed vertical axis $\mathbf{K}$. Therefore the student writes

$$\boldsymbol{\omega} = \omega_1 \mathbf{i} + \omega_2 \mathbf{j} + \omega_3 \mathbf{k}$$

and

$$\boldsymbol{\Omega} = \Omega \mathbf{k}.$$

Under gravity and neglecting friction the only external force acting is gravity, so the external torque is

$$a\mathbf{k} \times (-mg\mathbf{K})$$

where a is the distance of the centre of mass of the top from the point about which it pivots, usually its contact point with the ground. The two axes $\mathbf{i}$ and $\mathbf{j}$ have not been assigned precisely other than that they are in the plane

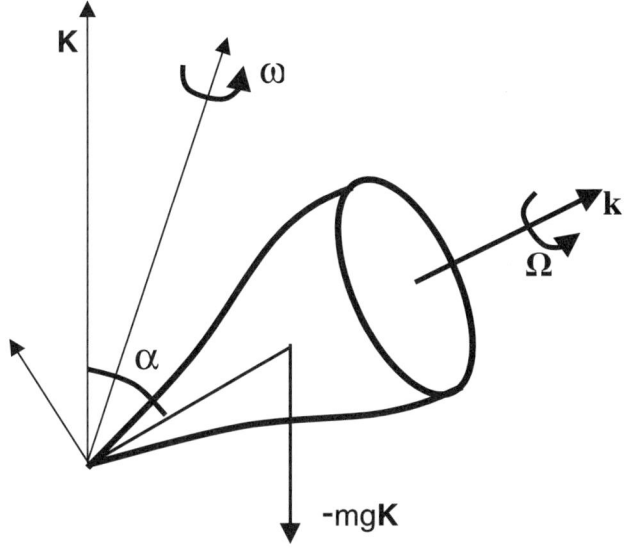

Figure 3.3   A spinning top.

of rotation of the top. Therefore it is possible to define **i** as pointing in the direction of the torque. Thus

$$\mathbf{K} = \sin\alpha\mathbf{i} + \cos\alpha\mathbf{j}$$

and

$$\boldsymbol{\Gamma} = mga\sin\alpha\,\mathbf{i},$$

where the letter $\boldsymbol{\Gamma}$ is standard notation for torque. Euler's equations of motion can then be written:

$$\begin{aligned}
A\dot{\omega}_1 + (C-A)\omega_2\omega_3 + C\omega_2\Omega &= mga\sin\alpha, \\
A\dot{\omega}_2 + (A-C)\omega_1\omega_3 - C\omega_1\Omega &= 0, \\
\dot{\omega}_3 + \dot{\Omega} &= 0,
\end{aligned}$$

where A and C denote the moments of inertia of the top about **i** (or **j**, it matters not which due to axisymmetry) and **k** respectively. By the way the axis **i** has been defined, and the assumption that α is constant and in the plane of **j** and **k**, $\omega_1 = 0$. The second of the above equations then gives $\dot{\omega}_2 = 0$ which implies ω_2 is a constant or that the top precesses steadily. This then provides a summary as it appears in this project of the background needed to understand precession which is essential for realising why a boomerang or frisbee flies like it does.

Apart from technical facility with vectorial mechanics, it is important for the student to understand the conservation laws of energy and angular momentum. The above vector algebra indeed shows good understanding. Of course the student can expect to be quizzed on this understanding at the oral examination, hence copying without understanding remains a high risk strategy to adopt. Under the assumption of a constant tilt of the axis of symmetry the precession must be steady. But if more general motion is to be understood, these conservation laws, in particular the conservation of angular momentum, can help to predict the behaviour of spinning flying bodies such as frisbees and boomerangs. As the scope of the project indicates, another important topic to understand is aerodynamics.

A number of pages in the project are devoted to the introduction of inviscid flow in two dimensions. This leads to the Joukowski transformation which is presented in the form

$$Z = z + \frac{c^2}{z}$$

where Z and z are two complex variables. The project student then proceeds to show that this transformation transforms a circle into an ellipse. Here are some details from the project. A circle in the z-plane of the form

$$z = ae^{i\theta} \text{ where } 0 \leq c \leq a$$

is considered. This is substituted into the Joukowski transformation to give

$$\begin{aligned} Z &= ae^{i\theta} + \frac{c^2}{a}e^{-i\theta} \\ &= a(\cos\theta + i\sin\theta) + \frac{c^2}{a}(\cos\theta - i\sin\theta) \\ &= \left(a + \frac{c^2}{a}\right)\cos\theta + i\left(a - \frac{c^2}{a}\right)\sin\theta \end{aligned}$$

which since $Z = X + iY$ gives

$$\frac{X^2}{\left[a + \frac{c^2}{a}\right]^2} + \frac{Y^2}{\left[a - \frac{c^2}{a}\right]^2} = 1,$$

that is, an ellipse in the Z-plane. The inverse of the Joukowski transformation is

$$z = \frac{1}{2}Z + \left[\frac{1}{4}Z^2 - c^2\right]^{1/2}$$

and this is substituted into the complex potential for flow (U) past a cylinder with circulation κ

$$w(z) = U\left[ze^{-i\alpha} + \frac{a^2}{z}e^{i\alpha}\right] - \frac{i\kappa}{2\pi}\ln z$$

to give the complex potential for flow past an ellipse with circulation. This particular student had to derive this potential in full: however this would not be done if two dimensional aerodynamics formed part of a module. Some courses contain thin aerofoil theory which is interesting mathematically. In such a university, comparisons with this approximate theory could be made, but not in Plymouth. Next, an off-centre circle $z = \gamma + ae^{i\theta}$ is transformed into an aerofoil from which lift is calculated. As it is derivatives of the potential w that give the flow and not w itself the calculation is not difficult:

$$w(z) = U\left[(z+\gamma)e^{-i\alpha} + \frac{(a+\gamma)^2}{z+\gamma}e^{i\alpha}\right] - \frac{i\kappa}{2\pi}\ln(z+\gamma)$$

so substituting for z through

$$z = \frac{1}{2}Z + \left[\frac{1}{4}Z^2 - a^2\right]^{1/2}$$

gives

$$\frac{dw}{dZ} = \frac{dw/dz}{dZ/dz} = \frac{U\left[e^{-i\alpha} - \left[\frac{a+\gamma}{z+\gamma}\right]^2 e^{i\alpha}\right] - \frac{i\kappa}{2\pi(z+\gamma)}}{\left[1 - \frac{a^2}{z^2}\right]}.$$

Interesting mathematical points about the flow at the singularities at $z = \pm a$ are discussed, and it is deduced that a circulation

$$\kappa = -4\pi U(a+\gamma)\sin\alpha$$

gives a flow free of singularities. The constant γ is assumed real in this project; the student overlooked the possibility that it might be complex. In fact a complex value gives a more realistic aerofoil, but some of the algebra can be confusing. Lift is approximately $\pi\rho U^2 L \sin\alpha$, where the circulation κ is neglected in comparison with a and L is the length of the wing (radius of the frisbee or "radius" of the boomerang defined from the geometry in some way).

The project then outlines the relationship between vortex shedding and lift before bringing together the mechanics and fluid dynamics to discuss the flight of frisbees and boomerangs. The boomerang is analysed and some attempt is made to understand why it returns to the thrower. The problem is a complex one and this part of the project is rather descriptive and not particularly convincing. So although there is good mathematics in the project ranging over solid and fluid mechanics, and an understanding of the practicalities of throwing frisbees and boomerangs, the project has shortcomings in terms of knitting the three parts together. There were also some mathematical errors and some parts where the student fails to be as general as possible. The project topic could be criticised as being rather too wide in scope. This is not entirely the fault of the student, although (s)he should perhaps have sought more guidance. The student who was responsible and mature in attitude although not in years certainly enjoyed the project and gained a lot from it.

It is sometimes difficult to define the scope of a project tightly enough at the beginning. One problem is that the mathematical experiences of the project supervisor at the equivalent age can be widely different to the student. This is not just a generation gap. In mechanics in particular, modern students in the UK lack the facility with algebra and geometry to get to grips quickly with new mechanics. This is true even for able students: it is as if the oil is missing from the bearings. It is all too easy to overestimate the student's ability and assign an over-ambitious project. This one in retrospect did err on the side of being over-ambitious although not disastrously so. Both assessors were in agreement over the mark to award this project: they both felt that the student was very strong at presentation and answering questions, but that there were some shortcomings in the student's understanding of the mathematical detail. The mark was between 60 and 65%.

3.6 Hurricane Dynamics

The general topic of geophysical fluid dynamics is a reasonably recent addition to applied mathematics. Although the ocean, atmosphere and inner earth have been described qualitatively since the seventeenth century (the inner earth rather inaccurately!), it is only in the twentieth century that any substantial quantitative modelling has been done. Modelling in this area is a young science and is still very active thanks largely to the availability of computing power together with resources arising from public interest in the environment. Hurricanes are one particular kind of severe tropical storm. They are publicly well known but continue to be very hard to model accurately. A hurricane is a complex system. It is a three dimensional swirling time dependent object with sophisticated fluid dynamics and thermodynamics. There is still much that is not known about hurricanes, so to do them justice in a final year mathematics project is difficult.

3.6.1 Scope

Faced with the above, the student on this project needed to make some decisions. The idea to do "something about hurricanes" came from the student, so the desire was there but a vague idea for a project has to be honed into shape quickly and this is one of the roles of the supervisor. After discussions the student on this project decided on a particular strategy and the introduction consisted of a description of a simple physical model together with diagrams. The first chapter of the project is mathematically very elementary and contains only a short description of ideal fluid mechanics and a general description of dynamic meteorology which is not really an introduction to the project but more of an introduction to the subject of meteorology. Instead of quoting this, here we give an excerpt from this student's chapter 2 which is an introduction to the hurricane.

> Figures 3.4 and 3.5 show a vertical cross section of the hurricane, which consists of four distinct regions and may be described as follows:

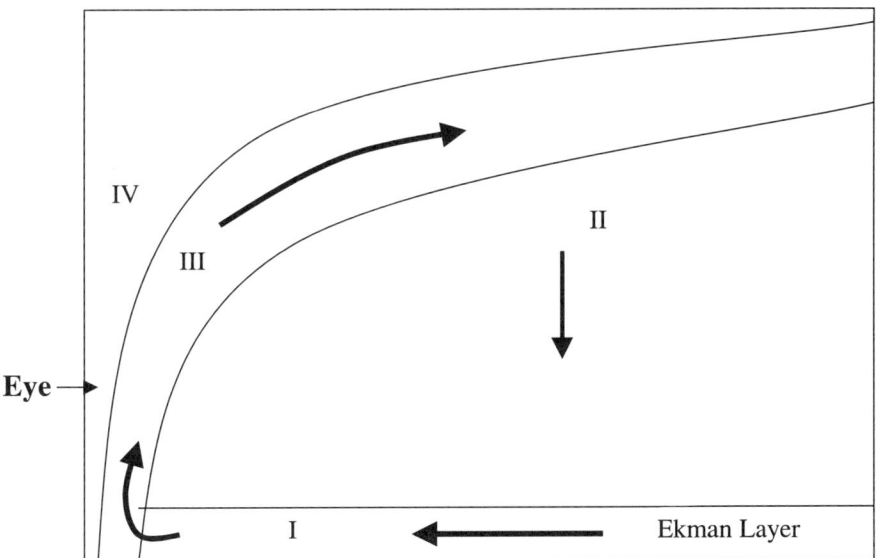

Figure 3.4   Schematic cross section of a hurricane.

- An updraft region of tall convective clouds and intense rainfall with accompanying release of large amounts of latent heat (region III)
- A region of rapidly swirling winds corresponding to a large pressure gradient (region II)
- A boundary layer (region I) where the swirl winds interact with the sea

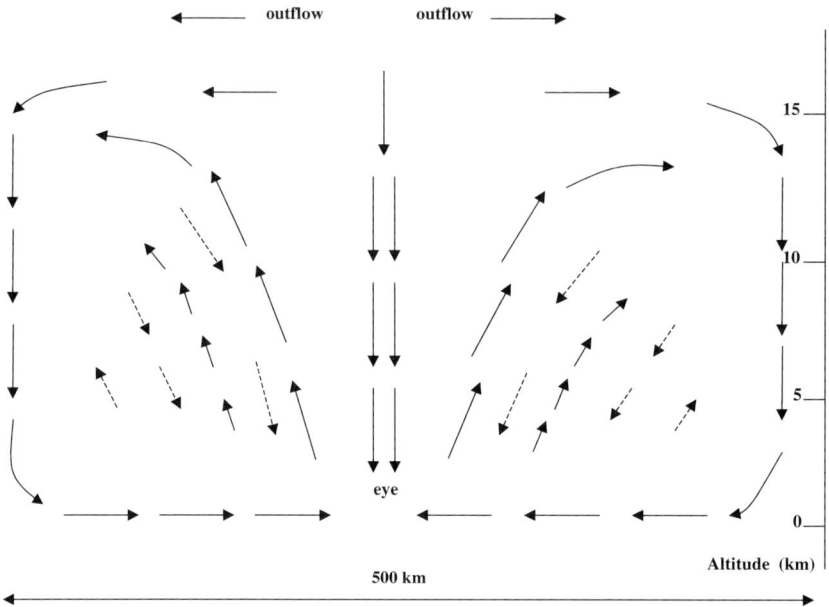

Figure 3.5   Cross section of a hurricane.

surface, causing an inflow which supplies moist air to the updraft in region III
- A warm quiescent, relatively dry core, or eye, in which there is a very slow recirculation. (region IV).

The swirling flow of region II settles very slowly into the region below, supplying the fluid, which moves radially inward in the boundary layer. There is almost no radial motion at all in region II and the flow is therefore effectively frictionless. The downdraught in region II is a source to the radial inflow in region I which moves into region III where it flows upward and outward, conserving angular momentum. At any given radial position, the updraft fluid has a deficit in angular momentum, compared to the fluid in region II. This is merely due to the frictional losses to the sea surface in the boundary layer.

Again note the absence of history (there was a brief sentence or two in the earlier chapter, but not enough to be classified as history). Instead we are straight into a model of a hurricane. This is typical of a project that concerns a model of a real situation. In the first few weeks of this particular project, the student was faced with two rather unpalatable facts. First it became apparent that the real hurricane was just too hard to model given the student's knowledge and experience. There could be no rainfall, no time dependent development and no lateral movement of the entire storm. The student was left to model a stationary rotating very idealised storm. Even so, the second fact was that there was some hard background theory of rotating fluids and thermodynamics which had to be understood. It is easy for the student to lose heart under these circumstances and the supervisor has to double as a counsellor. Here then are the agreed bullet points that cover the tasks of this particular student project:

- to understand geostrophic balance and thermal wind equations and solve them,

- to understand the Ekman boundary layer theory and apply it to a model of a hurricane,

- to be able to include enough basic thermodynamics to provide a convincing model of a hurricane.

The student on this project was very knowledgeable about hurricanes and possessed good background literature in terms of books about them. These texts contained little mathematics and were probably aimed at physical geographers. Dynamic Meteorology texts, e.g J. R. Holton's *An Introduction to Dynamic Meteorology* contained much of the modelling. However these are graduate level texts and were difficult for the student. One positive point was preparing and delivering the talk which was easy for this student. The talk itself consisted of a good general introduction to hurricanes together with the mathematics of geostrophy and Ekman dynamics.

3.6.2 Project Details

The subject matter here demands a familiarity with partial differential equations at a level a little above that normally expected of an undergraduate mathematics student at the start of the final year. However, the student did partially derive the simplified linear equations of the form

$$\frac{\partial u}{\partial t} - fv = -\frac{1}{\rho}\frac{\partial p}{\partial x} + F_x$$
$$\frac{\partial v}{\partial t} + fu = -\frac{1}{\rho}\frac{\partial p}{\partial y} + F_y$$
$$0 = -\frac{1}{\rho}\frac{\partial p}{\partial z} - g + F_z$$

the first two of which express geostrophic balance and the last of which expresses hydrostatic balance (provided the time dependence is ignored and the externally applied force $(F_x, F_y, F_z) = \mathbf{0}$). The notation, standard for meteorology and oceanography, is as follows: local Cartesian co-ordinates (x, y, z)

are x-East, y-North and z-up with associated velocity vector $\mathbf{V} = (u, v, w)$. (w does not feature as it is so small compared with u and v.) p is pressure, ρ is density (constant in the first instance), $f = 2\Omega \sin \theta$ is twice the vertical component of the Coriolis acceleration usually called the Coriolis parameter (Ω is the angular velocity of the earth, θ is latitude), and of course t is time. The student learned the consequences of these linear equations. For example that in the Northern hemisphere wind is parallel to the isobars and travels around a low pressure centre (and a hurricane) anti-clockwise in the absence of friction. The student also realised that friction causes the wind to turn inwards towards the low pressure giving the characteristic spiralling pattern of clouds seen in satellite photographs. The natural co-ordinate system to use when modelling a hurricane is plane polar co-ordinates. However there are problems with this, for as the student soon found, unfortunately the Ekman equations (given here later) are insoluble in polar co-ordinates. The quantity $\zeta = \nabla \times \mathbf{V}$ is the vorticity which as the student soon found out is an important quantity to understand, especially the vertical component of this added to f which is conserved as the hurricane moves as a whole (the conservation of potential or absolute vorticity).

At the ground friction becomes important to include in a model. The most successful model, now nearly one hundred years old, is due to Ekman and assumes a boundary layer in which a constant eddy viscosity parametrises the complex turbulence and skin friction. This is the model that the student used. The first two of the above equations without the time dependence become

$$-fv = -\frac{1}{\rho}\frac{\partial p}{\partial x} + \nu \frac{\partial^2 u}{\partial z^2}$$
$$fu = -\frac{1}{\rho}\frac{\partial p}{\partial y} + \nu \frac{\partial^2 v}{\partial z^2}$$

where ν is the eddy viscosity. Given the linearity of these equations, the student has little difficulty with the following procedure for solving them. First of all the geostrophic part of the horizontal velocity is separated from the rest and given the suffix g so that $(u, v) = (u_g + u_E, v_g + v_E)$ and

$$-fv_E = \nu \frac{\partial^2 u_E}{\partial z^2}$$
$$fu_E = \nu \frac{\partial^2 v_E}{\partial z^2}.$$

The complex velocity $w = u_E + iv_E$ is introduced and the first equation added to i times the second to give

$$\frac{d^2 w}{dz^2} + \frac{if}{\nu} w = 0$$

which is solved together with boundary conditions (zero velocity at the ground and geostrophic flow above the Ekman layer) to give the following solution:

$$u = u_g + u_E = -u_g e^{-\gamma z} \cos \gamma z$$
$$v = v_g + v_E = v_g(1 + e^{-\gamma z} \sin \gamma z)$$

where $\gamma = \sqrt{f/(2\nu)}$ and it has been assumed that the x-axis has been aligned to the geostrophic wind (i.e. $v_g = 0$). The difficulty the student has with this part of the project is not so much the mathematics (although there are notation problems, not reproduced here to avoid unnecessary confusion) but the interpretation of the physics. The formulation of a boundary value problem out of the assumptions made and then interpreting the solution is unfamiliar territory for many students. Incidentally, this Ekman model resurfaces (no pun intended) in Chapter 5 as an example of a case study, see Section 5.2. This is deliberate and enables direct comparison to be made between an individual project and a case study when the background topic is the same.

The above is mathematically tidy. However, modelling hurricanes is not done using neat mathematics and the rest of the report wrestles with some mathematical models which do not have tidy solutions. The student found immediately that if one included the curvature of the isobars of a hurricane this produces an extra term $-v^2/r$ on the left hand side of the x equation which stops the polar version of the Ekman equations from being solved exactly. The student therefore took the Cartesian solution (Ekman spiral) and inferred correctly that a rotary geostrophic motion must cause inflow towards the eye. From this a general picture begins to emerge, although there is much work to be done before anything like a workable model of a hurricane takes shape. The middle chapters of the project are devoted to atmospheric thermodynamics and a description of the fluid flow in the eye of the hurricane. The first and second laws of thermodynamics are given in mathematical form and the project here looks sophisticated. For this description we omit the details, but for those not versed in the subject its principal complexity lies in the plethora of symbols and definitions as well as keeping track of what is constant and what is varying when the partial derivatives are calculated. Here is one example from the project. The first law of thermodynamics in a meteorological context can be expressed as

$$\theta = T \left(\frac{1000}{p}\right)^{R_d/c_p}$$

where θ is the potential temperature, T is the temperature, R_d is the gas constant for dry air and c_p is the specific heat for dry air at constant pressure. It can be seen that this is assuming dry air and ideal gas laws. It can also be seen that there are terms that need definition, and these definitions in turn lead to the introduction of new concepts. The student was faced with a very real problem of how much to include. In the end, not enough explanation was given to the liking of one of the assessors and the student lost marks because of this. This student probably included too much mathematical detail and gave the assessors too much rope with which to hang the unfortunate undergraduate. The next bit of mathematics shows this. The expression

$$c_p T = \ln\left(\frac{\theta_{es}}{\theta}\right) = L_c q_{es}$$

was derived from the last equation courtesy of three definitions, saturation equivalent potential temperature, the latent heat of condensation and the saturation mixing ratio. This equation is then differentiated and the saturated moist entropy $\hat{S}$ is defined which enables the expression

$$c_p T d(\ln \theta_{es}) \approx T d\hat{S}$$

to be derived. There is then much more partial differentiation and introduction of definitions, all of which are correct, but not all of which seem to have been understood by the student.

Combining this complicated thermodynamics with the hydrodynamics of a swirling storm was then attempted with partial success. The last chapter re-introduced the time dependent terms in the Ekman equations and Laplace transforms were used to solve them. The solution to the time dependent problem in s-space (where s is the Laplace transform variable) is given as

$$\Phi = \frac{\tau\sqrt{\nu}}{s\sqrt{s+if}} e^{-z\sqrt{\frac{s+if}{\nu}}}$$

where Φ is the Laplace transform of the complex velocity $u + iv$. The inversion of this was not possible, but final and initial value theorems were used to good effect to deduce the time dependent behaviour and the student re-derived some results which had only previously been given in integral equation form (Fredholm Integral of the first kind) by use of the convolution integral. At this point this project showed some similarities to the boomerang project of the last section in that the pieces did not hang well together. Some differences of opinion between assessors also surfaced. The supervisor realised the difficulty of the project and wanted to award a mark just over 60%. However, the second assessor was less impressed. All he saw was a rather disjointed and bitty project, albeit on a difficult subject, but crucially during the oral examination the student seemed unable to answer any penetrative questions. This left the second assessor with the impression that much of what was written was not at all well understood. Eventually a compromise mark of 56% was reached. This student was rather disappointed with this mark, but the lesson is as follows:

better to reduce the scope of the project and really understand all that is written than be too ambitious, even when deep personal interest makes the latter extremely tempting.

These lessons seem to be well understood by the student who took on the following project.

3.7 Hypergeometric Functions

Hypergeometric functions are solutions to a second order differential equation. They used to feature widely in the final stages of mathematics courses, but do so rarely these days having largely been displaced by numerical methods. However, students who enjoy solving differential equations see the hypergeometric equation as bringing together and generalising the solutions to more specific differential equations. Most students meet Bessel functions and Legendre functions, and solve these using series (the Frobenius method). Classifying the singularities of a differential equation and using the Riemann P function also feature in many degree courses. The student who pursued this project was particularly keen and quite happy to read and work through long passages of the classical work *Ordinary Differential Equations* by E. L. Ince originally published in 1926. The project starts with defining the hypergeometric equation

$$x(1-x)\frac{d^2y}{dx^2} + \{\gamma - (1+\alpha+\beta)x\}\frac{dy}{dx} - \alpha\beta y = 0,$$

finding and classifying the singularities of the equation then expressing the solution in the form of a Riemann P function for the hypergeometric function as follows:

$$y = P \left\{ \begin{matrix} 0 & \infty & 1 \\ 0 & \alpha & 0 \\ 1-\gamma & \beta & \gamma-\alpha-\beta \end{matrix} \; x \right\}.$$

In order to calculate the Frobenius series solution which was an application of a known method, the student had to know about and use Gamma and Beta functions. The series was then obtained. Special cases and different forms of the hypergeometric function were explored including the use of complex analysis. In the final chapter contour integral representations were found. Throughout the project, the technical level of the mathematics was high and there are several pages of complicated algebra that contain virtually no English words other than "therefore" and "hence"! However, the mathematics here is not new and the results are all to be found in texts on special functions or advanced differential equations. Indeed, the student found one or two very specialist texts devoted entirely to the generalised hypergeometric function. This project was technically demanding but the material was freely available. The scope of the project was not difficult to define. The student could stop after a particular task (say finding the series solution to the generalised hypergeometric function) or go on to something new (integral representation). Provided the student was technically competent to deal with the material and enjoyed tackling pages of algebra, this was a project that could earn high marks.

3.7.1 Scope

In a topic such as this, some history is certainly possible, but this student decided on a technical approach preferring to give some non-historical background on differential equations. Here is an excerpt from the introduction:

> The hypergeometric equation is a second order ordinary differential equation of a special kind. It can be shown, for example, that any second order O D E with regular singularities (see definition, Chapter 2) one of which is at infinity can be transformed into a hypergeometric equation. We show that this is so in some particular cases, but begin in Chapter 2 with finding the Frobenius series solution to the equation, generalising the series and examining the fact that special values substituted into the series yield some familiar elementary functions.
>
> By merging two of the singularities in a certain way we obtain what is called the confluent hypergeometric function. This is another species of hypergeometric function which can be further transformed into Whittaker's confluent hypergeometric function–in the form $w''(x) + p(x)w(x) = 0$, and also to the more familiar Bessel's equation.

It is not difficult to see from this short excerpt that the project soon gets into some intense mathematics. One can almost detect the impatience of the student to dispense with the words and get down to serious mathematics! It is important to stress to such students the vital role the individual project plays in providing a vehicle for the student to express complicated mathematics to a lay audience. This student found this particular aspect hard (although after graduating (s)he found work in the financial sector and is doing very well). The following bullet points were decided upon to define the scope of the project:

- to define and find the singular points and series for the solution to the hypergeometric equation,

- to relate special cases to well known elementary functions with proofs,

- to define the confluent hypergeometric function and examine its properties,

- to find integral representations of the hypergeometric function.

One avenue that was possible but not explored was to pursue a numerical approach and use computer programming to evaluate hypergeometric functions. One imagines this as a whole new project possibly suited to a student on a joint mathematics and computing degree.

As already mentioned there are plenty of texts available on hypergeometric functions, starting with the classical one *Modern Analysis* by Whittaker and Watson first published in 1902. The student found many texts including two entirely devoted to hypergeometric functions. The topic of the talk was a sample proof of a result in the project together with an overview of hypergeometric functions. At the time of the talk (halfway through the project) the student was not sure how much would be done, so there was a vagueness about the scope. The talk was technically good if a little dry. Also some students in the audience were a little overwhelmed by the mathematical detail.

3.7.2 Project Details

Of all the project examples given in this chapter, this one contains the most equations and the least writing, so to get a flavour it is necessary to plunge straight away into mathematical detail. The series solution to the hypergeometric equation is tedious but reasonably straightforward. One substitutes the series

$$y = \sum_{n=0}^{\infty} a_n x^{n+r}$$

into the differential equation

$$x(1-x)\frac{d^2y}{dx^2} + \{\gamma - (1+\alpha+\beta)x\}\frac{dy}{dx} - \alpha\beta y = 0$$

and equates powers of x. The lowest power of x leads to the indicial equation

$$r(r-1)a_0 + \gamma r a_0 = 0$$

from which $r = 0$ or $r = 1 - \gamma$. Straight away there is a special case $\gamma = 1$ to consider. The general coefficient of x leads to the recurrence

$$a_n = \frac{r^2 + n^2 - 2r - 2n + 2rn + 1 + \alpha r + \alpha n - \alpha + \beta r + \beta n - \beta + \alpha \beta}{(r+n)(r+\gamma+n-1)} a_{n-1}.$$

To get to this requires stamina, accuracy and knowhow to spot special cases for the parameters α, β and γ. All of which means mathematical ability. The digestion of all the algebra is considerably eased by the introduction of the Pochhammer symbol $(\alpha)_n$ defined by

$$(\alpha)_n = \alpha(\alpha+1)(\alpha+2)\ldots\ldots(\alpha+n-1) = \frac{\Gamma(\alpha+n)}{\Gamma(\alpha)}.$$

The student goes on to derive the series corresponding to the root $r = 0$ taking $a_0 = 1$ as

$$y = \sum_{n=0}^{\infty} \frac{\Gamma(n+\alpha)\Gamma(n+\beta)\Gamma(\gamma)}{\Gamma(n+\gamma)\Gamma(\alpha)\Gamma(\beta)n!} x^n = {}_2F_1(\alpha, \beta, \gamma; x)$$

using the standard notation ${}_2F_1(\alpha, \beta; \gamma; x)$ for the hypergeometric function. The second solution corresponding to $r = 1 - \gamma$ is derived similarly:

$$y = x^{1-\gamma} {}_2F_1(\alpha - \gamma + 1, \beta - \gamma + 1; 2 - \gamma; x) = \sum_{n=0}^{\infty} \frac{(\alpha - \gamma)_{n+1}(\beta - \gamma)_{n+1}}{(-\gamma)_{n+2} n!} x^{n-\gamma+1}.$$

Obviously when $\gamma = 1$ the two solutions are no longer independent and a second solution has to be sought through letting

$$y_2(x) = y_1(x) \ln(x) + \sum_{n=0}^{\infty} c_n x^n$$

where $y_1(x) = {}_2F_1(\alpha, \beta; \gamma; x)$ is the first solution. This is standard procedure in the Frobenius method but still leads to lengthy algebra which needs to be done accurately. The books are content with "it can be shown that..." so copying a textbook is not an easy option.

There are proofs of the following relations:

$${}_2F_1(-N, c; c; -x) = (1+x)^N,$$

$${}_2F_1(\frac{1}{2}, 1; \frac{3}{2}; -x^2) = \frac{\tan^{-1} x}{x},$$

$$\lim_{c \to \infty} {}_2F_1(1, c; 1; \frac{x}{c}) = e^x$$

and

$${}_2F_1(1, 1; 2; -x) = \frac{\ln(1+x)}{x}.$$

The student obviously delighted in spending much time on the details of deriving these results, but they are straightforward. Less so is to show that Legendre's differential equation has solution

$$(x^2 - 1)^{\frac{1}{2}\mu}(A {}_2F_1(\frac{\nu + \mu + 1}{2}, \frac{\mu - \nu}{2}; \frac{1}{2}; x) + B {}_2F_1(\frac{\nu + \mu}{2} + 1, \mu - \nu + 1; \frac{3}{2}; x))$$

which the student also derived. The project also contains some reasonably genuine applied mathematics. There is a solution to the non-linear pendulum problem whereby the equation

$$\frac{d^2\theta}{dt^2} + \frac{g}{L} \sin \theta = 0$$

is derived where L is the length of the pendulum and $g = 9.81 \text{ms}^{-2}$. It is integrated once without difficulty and the solution is written

$$\sqrt{\frac{2g}{L}} \int dt = \int \frac{d\theta}{(\cos\theta - \cos\theta_0)^{1/2}}$$

which is expressed in terms of the elliptic integral of the first kind

$$F(k) = \int_0^{\pi/2} \frac{d\phi}{\sqrt{1 - k^2 \sin^2 \phi}}.$$

It is shown in the project that this can be written

$$F(k) = \frac{\pi}{2} {}_2F_1(\frac{1}{2}, \frac{1}{2}; 1; k^2)$$

and the period of oscillation is

$$T = 2\pi \sqrt{\frac{L}{g}} {}_2F_1(\frac{1}{2}, \frac{1}{2}; 1; \sin^2\left(\frac{\theta_0}{2}\right))$$

which becomes

$$T = 2\pi \sqrt{\frac{L}{g}}$$

when θ_0 is small as expected. The second piece of applied mathematics comes via the confluent hypergeometric equation. The original full hypergeometric equation is divided by β and β is allowed to tend to infinity. This has the effect of merging the two regular singularities at 1 and ∞ to a single irregular singularity at ∞. The differential equation becomes the simpler looking

$$x\frac{d^2y}{dx^2} + (\gamma - x)\frac{dy}{dx} - \alpha y = 0.$$

The solution of this is written

$${}_1F_1(\alpha; \gamma; x) = \sum_{n=0}^{\infty} \frac{(\alpha)_n x^n}{(\gamma)_n n!}.$$

This is derived in the project as is the solution to Bessel's equation

$$x^2 \frac{d^2u}{dx^2} + x\frac{du}{dx} + (x^2 - n^2)u = 0$$

in the form

$$u(x) = Ae^{-ix} x^n {}_1F_1(n + \frac{1}{2}; 2n + 1; 2ix) + Be^{-ix} x^{-n} {}_1F_1(-n + \frac{1}{2}; -2n + 1; 2ix),$$

and a special polynomial solution to Laguerre's equation

$$x\frac{d^2y}{dx^2} + (1-x)\frac{dy}{dx} + Ny = 0$$

in the form

$$L_N(x) = {}_1F_1(-N;1;x) = e^x \frac{d^N}{dx^N}(x^N e^{-x})$$

where N is an integer. It is these Laguerre polynomials that feature in mathematical models of edge waves in the ocean. This too was mentioned in the project but without a lot of detail. In a project such as this where there is an emphasis on mathematical methods it is good to at least nod in the direction of some applications. Another route could have been to relate hypergeometric functions to the solution of Mathieu's equation which is found both in jet engine outflow modelling and the modelling of sub-harmonic oscillations, but this was not done here.

The final chapter of the project contained the derivation of integral representations of the hypergeometric functions. The student knew complex variable theory up to the residue calculus, but had not met branch points before. The properties of these had to be learned, for example using branch cuts to render a complex plane single valued. There were more pages of mathematics, this time complex contour integration with branch cuts and small circles the radii of which are made infinitesimally small before the result

$$_2F_1(\alpha,\beta;\gamma;z) = \frac{\Gamma(\gamma)}{2\pi i \Gamma(\alpha)\Gamma(\beta)} \int_{-\infty}^{\infty} \frac{\Gamma(\alpha+s)\Gamma(\beta+s)\Gamma(-s)(-z)^s}{\Gamma(\gamma+s)} ds$$

is derived and shown to be convergent for $|z| < 1$ and $|\arg(-z)| < \pi$. The contour is a semi-circle in the complex s plane and (of course) z is an arbitrary complex number.

In a project such as this, it is important for the student to be analytically correct. The convergence of infinite integrals, the range of differentiability and continuity of functions need to be stated, and the student who wrote this project was careful in this regard. What emerged here was a carefully and accurately done project which contained difficult mathematics. All the mathematics (apart from the edge wave application barely mentioned) was nineteenth century but a great deal of it was new to the student. The topic had the advantage of being "well known" in the sense of being embedded in the literature. Although some literature sources were difficult to follow there was no question as to the correctness of the mathematics. The student had to work harder to explain what (s)he was doing to an intelligent lay audience. If the project as a whole had faults, they lay in this direction. The assessors were impressed with the project report, and the student performed very well in the oral examination, answering all questions clearly, confidently and, most important, correctly. Overall the student did very well on this project, and as you may have already guessed was unanimously awarded a clear first.

3.8 Summary

When mathematics projects first started in the early 1970s in the UK, it was thought they would be on topics such as "ten ways to calculate π". The hypergeometric project here is a direct legacy to this kind of project in mathematics. It is a skills based project of the type commonly found in mathematics, engineering and the physical sciences, but absent from the social and political sciences and business studies. The other three are in a sense more mature in structure. The first is really embedded in the history of mathematics, geometry in particular, and the student has the opportunity to be scholarly, read around the subject and the end result could almost be publishable. The boomerang project was about modelling the behaviour of a device. This meant selecting the appropriate mathematics from the student's store and modelling the motion. This is classical mathematical modelling with the emphasis being on the model building and less on verification. The hurricane project is more difficult. It attempts to describe mathematically a natural phenomenon. This is harder than describing boomerangs for several reasons, not least because hurricanes cannot be controlled. Again the emphasis is on building the mathematical model of the hurricane rather than on verifying its accuracy precisely. In both of these projects, the students knew the shortcomings of their models. They both learned the value of mathematics to model real world phenomena which is a definite plus. The geometry project gave the student the opportunity to write in a scholarly fashion about a highly mathematical topic. This too is a plus. The more traditional hypergeometric equation project enabled the student to hone mathematical skills (algebraic manipulation and analysis) to a substantial depth and a depth not possible in taught modules, and this is a plus. All the projects were therefore successful. There are many other project topics that might by many be considered distinctive and unlike those outlined here, but I would disagree. Only a project based on computing might perhaps be different. If it involved solving a real problem numerically it would resemble the boomerang project in form, and if it were more "pure" computing (logic or formal methods) it could resemble the hypergeometric equation project. There may be a hybrid (encryption perhaps) that could have different features, but these features would not be significantly distinctive. There is therefore certainly enough material in this chapter to help to see how a range of individual projects in mathematics operate.

The assessment of individual projects was outlined in Chapter 2 alongside assessment in general. The main part of the assessment has to be the project write up. It is normal for this to be say 80% of the marks with 10% each for the oral assessment (viva-voce or alternative) at the end and a further 10% for the verbal presentation midway through. Sometimes an interim report is asked for (see Chapter 4) and marks are assigned to this. This report should be short, less than 5 pages, and give an indication of what has been achieved so far and a view of what is to come. The 80% of marks for the written project is often subdivided further either explicitly or in the form of guidelines. I prefer the latter for maximum flexibility. The project write up should be assessed for *content* and *presentation*. Content is principally the level and accuracy of the mathematics, but is also the correctness and standard of the written English. Presentation covers the layout, readability, literary style and general appearance of the finished article. The assessors, normally two in number, usually meet once all the various bits of assessment are assembled and then they agree over the final mark. If there is a dispute, a third assessor needs to act as referee. In my experience, where there have been disputes it is almost always the supervisor doubling as first assessor being more generous than the second assessor and again almost always a compromise is quickly reached, usually because the second assessor is made more aware of some of the circumstances.

Normally a chapter in a mathematics text ends with a set of exercises. This cannot be done sensibly here so instead a list of possible project titles is given together with what hopefully are useful comments.

3.9 Project Examples

In all the projects that follow, a student has successfully followed through an investigation and has gained an honours degree either in Mathematics, or with mathematics as the major subject (i.e forming two thirds of the programme of study). No statistics or statistics related projects have been listed.

The Numerical Solution of Ordinary Differential Equations using Adaptive Step-length Techniques

This project is based squarely in numerical methods and extends the student's experience by solving ODEs using methods not met in the classroom. The project can contain a review of those techniques met in class such as Runge-Kutta and some multi-step methods, but will move quickly on to other multi-step methods and the use of a variable step-length to control the truncation

3. Individual Projects

error. There is also the opportunity to write computer programs, or to understand and perhaps adapt existing software.

Set Theory

Here is an example of a project based in Pure Mathematics. This project concerns ZFS (Zermolo, Fraenkel and Skolem). It needs to contain a discussion of ordinal and cardinal numbers, but ultimately it is about the Axiom of Choice: given any collection of non-empty sets A, B, C,... there exists a set consisting of exactly one member from each of the sets A, B, C,.... Of course, the content and structure of this and any of the projects mentioned here will depend on the scope of the modules the student is taking.

Braid Groups

Here is another Pure Mathematics project. However it is not as pure as the one above. Braids are topological objects linked (no pun intended) to knot theory but with applications to areas of geometry, dynamic systems and physics. There are many possible projects, but the most natural direction to take would be an extension to a final year pure mathematics module on topology looking at the group structure and equivalence classes (Artin's Theorem). There is plenty of opportunity to prove relatively simple theorems in a subject that lends itself to practical visualisation.

Population Dynamics in Moths

As the title implies, this is a project based around solving a set of differential equations that emerge from a population model. There could be material on the stability or otherwise of such sets of equations, but only if they are not included in the course. Extensions to Liapunov criteria and Hopf bifurcation can be made, and there is the opportunity to validate the model using real data. (If there is data on other populations, perhaps the Canadian hare and lynx, then the title of the project can be adapted accordingly).

Fundamental Concepts in Quantum Computing

This is a relatively new subject, but *may* provide the basis for efficient computational algorithms and unbreakable codes. However, the student needs to start by understanding the Dirac notation and relating quantum concepts such as spin to logic gates. If the student has no experience of quantum mechanics and no experience of logic then perhaps this is not for him/her. The potential of quantum computing to save vast amounts of money by making networks much more efficient is tempting big companies to invest heavily into research. It is therefore a good area for students to get into.

The Mathematics of Sound Wave Motion

This project derives the wave equation for sound waves and investigates its properties. If there is a module on waves on strings, then the derivation can be omitted, but there is still plenty of scope to look at modification due to porosity, incomplete reflections, etc. Investigation into cylindrical waves and spherical waves lead to Bessel and Legendre functions, so the student can play fun and games with these. Another possibility is the use of mathematical software such as Maple to display the waves, especially the three dimensional waves. This topic is large and it is not difficult to think of many projects.

The Three Classical Problems

These problems are: doubling the cube; squaring the circle; trisecting the angle. This project suits a student who likes a well defined project and likes more or less to be told what to do. Having said that however, there is scope here for a lot of abstract algebra in the form of extension fields. The student can prove various results and provide examples and proofs before delivering the *coup de grace* by using extension fields to show that all three problems are impossible to solve. This project has to be done from the point of view of rigorous mathematics of course: a cynical applied mathematician could always draw a square of side $\sqrt{\pi}$ to duplicate the area of a unit circle, or manufacture a cube of side $\sqrt[3]{2}$ to double the volume of the unit cube!

The Quark Model

This project is most definitely theoretical physics. The Lie group has to be introduced, and hence some background on groups is essential. Representations are defined and the group $SU(2)$ shown to be useful, but not adequate and $SU(3)$ is required to represent the quark. This project requires the facility to work with definitions and mappings as well as some manipulative ability with 3×3 matrices. It is the sort of project that may tempt the weak student because it sounds interesting, but the temptation should be resisted.

Trapped Waves

This project involves solving the linear shallow water equations with rotation in order to find the special waves that are trapped to travel along a coast. The Kelvin wave is the classical solution, although it is also possible to derive equations that describe the Continental Shelf Wave. These latter waves do not need a free surface elevation but demand a depth change. Relating Kelvin waves to tidal oscillations in specific geographical locations such as the North Sea can be a rewarding validation experience for the student. This project demands knowledge of inviscid fluid mechanics and some elementary differential equation theory. However, by far the most challenging aspect of all projects that try to map the real world is the modelling: what to leave in and what to delete from all the choices. For final year undergraduates, almost always it is the supervisor that has to make the choice.

Bessel Functions

This is another project which is very tightly defined. The student is given Watson's big book on Bessel Functions and told to go away and understand it! Well, not quite, but series solutions, special cases, Fourier-Bessel series and integral representations can all get an airing within such a project. It is more than likely that Bessel's equation has been met on the course in one form or another, in which case duplication must be avoided. The use of complex analysis to explore the properties of the equation and the different kinds of Bessel function is another profitable avenue for some students to take. As well as the standard $J_\nu(z)$ and $Y_\nu(z)$ there are the modified Bessel functions $I_\nu(z)$ and $K_\nu(z)$ and the Hankel function $H_\nu = J_\nu + iY_\nu$. Then there are the spherical Bessel or Kelvin functions $ber(z)$ and $bei(z)$, $ker(z)$ and $kei(z)$, and there is the Whittaker function and links to the confluent hypergeometric function to explore. Plenty of scope here, but the student must be prepared to do some manipulative algebra and calculus. It resembles the hypergeometric function project outlined in Section 3.7.

The Magnetic Field of the Milky Way Galaxy

This project is undeniably astronomy, and obviously suits the student whose interests lie in this direction. Some knowledge of astrophysics is certainly desirable in order to understand what the field is, as well as concepts such as the polarisation of starlight and Faraday rotation. However, the actual mathematics involved needs to be little more than trigonometry, and a little bit of mechanics. The hard part is understanding what this represents which makes this more of an astronomy project than a mathematics project. As long as there is agreement about this (perhaps the student is studying for joint honours) then there should be no problem.

Let us finish by listing some further project titles that have been successfully pursued. The first two are joint Mathematics with Education student projects and the same comments as made above about astronomy would apply.

Table 3.2. A project list

Title
An investigation into the implementation of the framework for teaching mathematics at key stage 3 (age 12-14)
Using case study methodology to determine what MTutor © must provide for student learning to be successful
An investigation into the implementation of the National Numeracy Strategy
The four colour problem
Rotating fluids
Protostellar genesis
The mathematical modelling of mechanical systems
Flow of a Newtonian fluid in an L-shaped channel using separation of variables and matching techniques
An introduction to chaos theory through a simple physical example
The Klein correspondence between twistor theory and space-time
Wave power devices
Dimensional regularisation in quantum field theory
Knot tabulation
The conservation of planetary vorticity and its applications

We have seen in this chapter how a wide variety of interesting topics are suitable vehicles for a project in mathematics.

Individual projects may still be a novelty to some professional academic mathematicians. Group projects in mathematics, the subject of the next chapter, are likely to be new to everyone.

4
Group Projects

4.1 Introduction

One distinctive feature of mathematics is that at undergraduate level there is virtually no opportunity for debates about different approaches or differences of opinion. This contrasts with most other subjects where the opportunity for debate is often central to the development of the student. For example, debating is the very essence of political science, and in the arts the ability of a student to be able to voice an opinion and back it up with argument is extremely important. Even in the sciences, although differences of opinion are rarer, there is always the opportunity to work together in the laboratory or to co-operate and work as a team towards a single goal as part of a large project. In the real world, one thinks of large teams working towards better treatment or even a cure for cancer and other dreadful ailments. The major achievements such as landing a man (person these days!) on the moon and returning him safely to earth, or being the first to climb Everest, or developing internet technology are all examples of successful team work. As discussion and debate are two of the primary objectives of group work one may legitimately ask what place this has in mathematics. Let us try to answer this. First of all there are undoubted social benefits. In a group, there can be a meeting of minds about a particular mathematical topic and this can help students learn from other students. A group can increase the self esteem of member students by helping when they get stuck on problems or by being sympathetic and otherwise supportive. Group work can really help the development of communication skills such as presenting an argument, conducting a meeting, learning to listen and respond, etc. All of these are key skills which mathematicians (amongst others) have been criticised for lacking on graduation. The acquisition of key skills is now very much a part of all education. These points hopefully provide reasons to try out one particular form of group work, the group project.

A group project consists of a group of two to six students working together to produce a piece of work which, as was said above, in mathematics is unusual. Of course, of all the disciplines, it is recognised that in mathematics it is possible to produce excellent work entirely on one's own. Some suggest that the best mathematics is produced by isolated individuals, but although this might be true for the most mathematically gifted, at undergraduate level most students benefit by not working in isolation. Even those students pursuing the individual project, the subject of the last chapter, are really doing so alongside their project supervisor as mentor and not by themselves.

In this chapter, some examples of group projects are analysed. The procedure of setting up group projects is outlined as well as management issues. Once this has been done, there are four examples of group projects. These examples are used to bring out what individual students do as well as the interplay (or otherwise!) between them.

4.2 Setting up Group Projects

At the outset, students need to know as precisely as possible what it means for them to be part of a group project. Obviously, at some point well before a decision needs to be made, information must be available outlining the way the group project is organised. A typical student handout was given in Chapter 1, and it is repeated here for convenience.

After this handout is distributed, more details of the assessment procedure is given. The peer assessment in particular intrigues most students. Details of the assessment are already given for the most part in Chapter 2. However it is worth going into particular detail here about the peer assessment.

4.2.1 Peer Assessment

When students work as a group, a new dimension enters assessment. If a group of students is working towards a common goal, then the final report has a single mark and all students who contribute to the report would gain the same mark. This is correct and represents collective responsibility for the final report.

4. Group Projects

Group Project

Organisation

A set of students, minimum number 2, maximum number 6, preferred number 3 - 5, work as a team on a project. This is called a Group Project.

At the beginning of the module, there will be a short sequence of lectures explaining each project. At the end of the lectures, students will be expected to choose one of them to work on. This does not preclude students from coming up with their own ideas for projects, but any ideas must be vetted for suitability by the module leader, and there must be a group of students willing to work together on it. In practice each project consists of a number of topics that individuals work on. There is a handout accompanying each project that includes suggested topics, however these are only suggestions and students working on the project are at liberty to formulate their own strategies for completing the project. Again, the module leader is there to help and provide guidance on, for example equalising workload within a project group. The goal is to produce a project report by the end of the period to which all the group has contributed more or less equally, and with which all members of the group is proud to be associated.

The module is worth 10 credits (there are 120 credits in the year) so a rough guide is that each student should spend 100 hours on their contribution.

After the initial lectures, formal class contact is restricted to one hour per week when the module leader is available in classroom for general consultancy, making general points from matters that arise, giving advice, etc. There are two other timetabled hours in the week when the module leader is available in his/her office. This should be used for specific queries, usually but not necessarily of a technical nature.

The choice of who does a particular project is largely up to you. You will be asked to state preferences, and these will form the basis of the split into groups. Note that it is perfectly possible for more than one group to work on the same project. Usually the projects have enough scope for many groups. The module leader does have the final say as it may be in your own interest to ensure that the composition of each group has roughly equal mathematical ability.

Assessment

The assessment of this module will be as follows:

1. Contribution to final report [hand in deadline]

   70%

2. Interim Report [hand in deadline]

   10%

3. Verbal report to the whole class [date of orals]

   10%

4. Peer assessment by other members of the group.

   10%

On the other hand, each student's contribution to the project as a whole is distinctive as is the view the rest of the group has on any particular student. How to assess this opinion is not easy. Here are a few suggestions, all of which I have tried.

1 One could simply use one's judgement and give each student a mark. This is very subjective and lays the module leader open to criticism.

2. A modification one might try is to interview each student and ask questions. Then armed with this information, together with one's own impressions, a mark is given to each student. The main problem with this remains subjectivity both on the part of the module leader and the students. It is open to bias especially if the students "gang up" on one particular student.

3. A different method is to give the group 100 marks and tell them to distribute these marks amongst themselves in an agreed fashion. The final mark given to each student is then this agreed mark scaled with the overall project mark. As an example, suppose there are four students with an agreed peer group mark distribution 40, 24, 20, 16. Say the overall mark for the group project final report was 50%. The students would then get 20, 12, 10 and 8 which if peer assessment was 10% of the total mark would contribute 2%, 1.2%, 1% and 0.8% to each student's mark respectively. I have used this method and the problem with it is that students tend to be kind to each other and agree a uniform distribution of marks. This was so even when it was obvious to everyone that in one particular case one student was being "carried" by everyone else. Another difficulty is that at extremes the scoring system fails. If a project gets 100% (or close) or 0% (or close) the module leader has to depart from the formula and give everyone the same excellent or awful mark.

In these days of quality audit and quality assessment, it is a good idea to have a second marker. This might seem impractical in peer assessment, but if one interprets this as having someone else oversee the process, then all of the above can be "second marked".

Most students like some idea of what is expected of them in terms of volume as well as content. In Plymouth, the individual project is worth 20 credits and should be 50 pages in length. In fact to discourage longer projects, the rule is that only the first 50 pages are marked (with some allowance for appendices, computer programs and the like). If the group project is 10 credits per student and the final report is worth 70% of the assessment then each student can expect to contribute about 18 pages ($\frac{7}{20} \times 50$, rounded up) and for a group of five the group project report can be expected to be about 90 pages long. Of course this is only a rough guide, but it is useful for students.

4.2.2 Dividing into Groups

Another question that concerns most students is how to select their peers. A method that seems to work well is to place on the notice board a chart. The body of the chart is empty, along the top are all the projects, and down the left hand side are the student's names. Each student is then asked to fill in the row alongside their name as follows. For the project they would most like put a one in the column, for the next preference put a two and so on until all columns are filled. If there are five projects from which to choose and there is a tie in preference, then instead of 1, 2, 3, 4, 5 there might be 1, 2, 2, 4, 4 or 1, 1, 1, 4, 5 in some order. The module leader then goes away and tries to devise a partition of the student body into groups that keeps everyone happy. Sometimes (quite frequently in fact) there is a reason other than an academic one for students to form particular groups. They may all live in the same flat or house or hall of residence, or they may all live in the same place but at some distance from the university which means they can meet more easily, especially out of term time. Perhaps they all share a hobby, maintain similar sleeping habits or are simply a group of friends who know they get on and want to work together. All these are perfectly good reasons for forming a group. However there is an argument for keeping the groups reasonably homogeneous as regards to mathematical ability. To have all the best (or all the worst!) students in the same group is less than ideal. I have concluded that it is preferable to have a happy group of students, so I let them choose and so far this has worked, with one or two exceptions.

So now we have groups of students who know in general terms what is expected of them, they have attended some preliminary lectures that outlined the projects available and have decided on a topic. They then chose a specific group project and know with whom they will have to work. Let us now see how they got on with particular projects.

4.3 Estuarial Diffusion

This particular group project involved five students. At the beginning of the semester, all the students were given some general lectures one of which was about modelling diffusion. Students were reminded about the technicalities of partial differentiation and how to solve differential equations using separation of variables (this technique for *partial* differential equations was new for most students). Students were told that they would meet techniques for solving partial differential equations including the use of Laplace Transforms in a module during the year should they choose it. Some information was then given about the physical process of diffusion in terms of the behaviour of molecules, together with the briefest introduction to modelling turbulence using eddy viscosity. Finally the students were reminded about finite difference methods and were told diagrammatically how these could be applied to partial differential equations of diffusion type. A brief mention of the underlying strategy behind the finite element method followed. Relevant concurrent lecture courses were also indicated. Following this, this particular group of students decided to do a project focusing on estuarial diffusion. This year, another group (of four) decided on heat transfer modelling as a topic. Therefore nine students chose to model diffusion, and rather than prohibit four from this general area two separate project teams were formed looking at these two different aspects of diffusion modelling. When one topic is particularly popular, I find that this is the best solution. One year, there were *four* groups on modelling analytical dynamics all doing different projects and we will meet one of these later in this chapter. It is still a mystery to me as to why one year a given subject is popular with students (this year diffusion) whereas another (this year dynamics) is unpopular, yet in the following year the reverse is true. It must be some kind of group dynamic as all other features such as the lecturer, handouts and mathematical experience of the student body are the same. The group project is a new and unfamiliar beast as far as mathematics students are concerned, therefore no matter how unfathomable their choices might seem, it is important to try and cater for them. However, to return to this particular project. A general handout on "Modelling Diffusion" accompanied the lecture, and this handout is reproduced here. Now, having read this handout and absorbing the lecture, the five students chose the following tasks. Student A researched the history of diffusion, writing some biographical details of A. Fick. The one dimensional diffusion equation was derived from first principles using Fick's assumptions. This extended and filled in the gaps from the lecture material. Quite a lot of physics was involved as well as historical research. This student did well. Student B was a capable mathematician and concentrated on finding analytical solutions to the one dimensional diffusion equation. For example, the separation of variables solution to the equation

$$\frac{\partial \phi}{\partial t} = \kappa \frac{\partial^2 \phi}{\partial x^2}$$

was found in the form

$$\phi = \sum_{n=1}^{\infty} Q_n \sin\left(\frac{n\pi x}{l}\right) e^{-(n\pi/l)^2 kt}$$

where the boundary conditions imposed are $\phi = 0$ at $x = 0, l$. The detail in the project was painstaking and other boundary conditions were considered such as

$$\frac{\partial \phi}{\partial x} = 0 \text{ at } x = l.$$

Perhaps the rest of the group found the mathematical rigour of this section a little frightening, but whatever the reason the solutions here were not related directly to estuaries in the other chapters. Student C examined numerical

4. Group Projects

Modelling Diffusion

1. Mathematical Nature
Analytical modelling using differential equations, techniques such as separation of variables, using Laplace transforms, specialist techniques such as using similarity variables. References: E. Zauderer *Partial Differential Equations of Applied Mathematics* 1989, H. F. Weinberger *A First Course in Partial Differential Equations* 1965. A specialist book on modelling diffusion is J Crank *The Mathematics of Diffusion* OUP 1975
Numerical Modelling using finite differences. References: Books on numerical solutions to partial differential equations e.g. G. D. Smith *Numerical Solution of Partial Differential Equations* OUP 1969.
Simulation using random walks and writing a computer program.
2. Possible Strategies
Start with the one dimensional diffusion equation

$$\frac{\partial \phi}{\partial t} = \kappa \frac{\partial^2 \phi}{\partial x^2}$$

where ϕ may represent temperature, x distance along a bar and t time. Investigate solutions to this equations using one or more of the methods listed under 1 above.
Look at more general equations:

$$\text{two dimensional} \quad \frac{\partial \phi}{\partial t} = \kappa \left(\frac{\partial^2 \phi}{\partial x^2} + \frac{\partial^2 \phi}{\partial y^2} \right)$$

and

$$\text{three dimensional} \quad \frac{\partial \phi}{\partial t} = \kappa \left(\frac{\partial^2 \phi}{\partial x^2} + \frac{\partial^2 \phi}{\partial y^2} + \frac{\partial^2 \phi}{\partial z^2} \right).$$

Investigate how the methods of solution generalise. What is κ and how is it determined? Investigate real life applications (e.g. diffusion in estuaries, see R Lewis *Dispersion in Estuaries and Coastal Waters*, John Wiley 1997)
3. Example of a Project Team Plan
Five members A, B, C, D, and E.

A Investigates one, two and three dimensional analytical solutions.

B Investigates numerical solutions.

C Investigates how κ is determined, solutions involving variable κ and alternatives.

D Investigates simulation using computing, e.g. particle tracking.

E Investigates extensions, e.g. non-linear modelling.

---

methods of solution. Once more, these were rather idealised situations and not directly related to estuaries. This student showed good understanding of finite difference methods and explicit, Crank–Nicolson and fully implicit solutions to the diffusion equation were analysed in terms of truncation error and stability. Finally this student applied a particle tracking random walk method (copied from a paper) to simulating pollution in an estuary. The fourth student D looked at non-linear diffusion. In particular the analytical solution to the diffusion of a contaminant in a river containing a stream of constant speed U. Laplace transforms are used, and many pages are covered with mathematics to derive that the concentration $c(x,t)$ that satisfies the equation

$$\frac{\partial c}{\partial t} + U\frac{\partial c}{\partial x} = \kappa\frac{\partial^2 c}{\partial x^2}$$

with $c(0,t) = c_0$, $c(\infty,t) = 0$ $t \geq 0$ and $c(x,0) = 0$ $x > 0$ takes the form

$$c(x,t) = \frac{1}{2}c_0 \left\{ \text{erfc}\left[\frac{x - Ut}{2\sqrt{\kappa t}}\right] + e^{Ux/\kappa}\text{erfc}\left[\frac{x + Ut}{2\sqrt{\kappa t}}\right] \right\}.$$

Finally, the fifth student E decided to concentrate on the phenomenon called the Kelvin–Helmholz instability. When two immiscible liquids flow over each other and the difference in the speeds is large enough then waves first appear then grow on the interface. In an estuary the two liquids are freshwater flowing over saline water, and this instability can be an important mixing mechanism which enhances diffusion. This mechanism for diffusion is different from those in the rest of the project as it is a form of *shear* diffusion caused generally where there is a sharp variation of current speed with vertical distance. By its very nature there was very little connection with the rest of the project; nevertheless the mathematics was of high quality. Water waves were analysed using linear theory and those aspects relevant to the discussion of the Kelvin–Helmholz instability highlighted. The student made use of the classic text *Water Waves* by J. J. Stoker 1957, the style of which is not easily accessible to undergraduates these days.

The outcomes of the theory are discussed in terms of the gradient Richardson number defined by

$$R_i = -g\frac{\partial \rho}{\partial z} \bigg/ \rho \left(\frac{\partial u}{\partial z}\right)^2.$$

The worthiness of this final contribution is not in question, but one wonders if the student would not have been happier pursuing an individual project.

About six weeks into the project, the individuals in this group submitted interim reports that showed each of them making reasonable progress. At least they all seemed to understand what they had to do which is the important thing. The reports were marked by the module leader and feedback was given. It is a good idea for a second assessor to get involved too. Perhaps they should second mark these reports. This was not done at this stage in Plymouth, but in subsequent years it was introduced as part of improvements in quality assurance processes.

The presentation to the whole class (about three months into the project) took the form of five mini-presentations. The student who was second up and tackled the analytical solutions overran which upset the timing for the three that followed. The style of presentation was fine if perhaps a little dry and over formal (due to nerves probably). The students clearly understood the mathematics that they presented and were able to answer questions well. Most were confident speakers which certainly helped.

The final report was submitted in time and the group seemed content with it. Some comments on this project report follow. In this project it is easy to see from the report that each chapter was written by a different student. There was no real attempt to link the chapters together apart from a sentence at the end of one or two of the type "In the next chapter we will discuss ...". If a serious attempt has been made to link the subject matter of each chapter then this is worth extra marks, but it is quite understandable for each student to do one chapter. Time is usually the critical factor, and in a project that only lasts one semester refinements such as rewriting chapters in the same style and swapping material around are editorial jobs that have a low priority. Another relevant point is that in this group, there was no elected leader who might have taken on such an editorial role. If the chapters were written in different typefaces and the equation numbering and paragraphs were different this would be less acceptable.

The peer assessment for this project was done by interviews. Each student was interviewed in isolation by the module leader. It came as no surprise to any of the students that all received the same high mark. If the students themselves had done the marking, the result would have been no different.

The feedback from this group of students was very positive and they clearly enjoyed and gained a great deal from working on the project. From the module leader's viewpoint on the whole the final written report was acceptable and represented the outcome of a successful project. The students certainly seemed to work well together. They happened to be all the same sex (female) and psychologists tell me that females are more co-operative than males who tend to exhibit rivalry in groups. My experience in running group projects agrees with this (so far). The module leader and an assessor each marked the written project report independently. Both gave it high marks, with the module leader's marks a little but not significantly above the independent assessor's marks. Both were of the opinion that the report represented the work of a group that worked well together and had produced a report of high calibre. The marks eventually agreed between the assessors were in the first class category and the external examiner was also in agreement with this. Two of the group went on to get first class degrees. (The other results were one upper second and two lower seconds.)

4.4 Graphs and Networks

Graphs have an interesting place in mathematics. A few years ago, they would be entirely absent from the mathematics curriculum. Even today, a lecture course on graphs would be seen to be unusual: moreover the syllabus would have to be scrutinised in detail to find out whether it was pure or applied mathematics. The reason for the recent upsurge in interest in graphs stems from the ability of computers to do large amounts of routine arithmetic. The kind of arithmetic required here is particularly suited to computers as it is essentially binary. Decisions have to be made whether to go along one path or an alternative subject to the satisfying of certain criteria. The study of graphs dates back to the prolific Swiss mathematician Leonhard Euler (1707–1783), but really only took off following advances in operational research during the Second World War. Now, there are whole industries using advanced networking algorithms particularly linked to the operation of the internet. Students therefore look upon an opportunity to work in this field with some keenness. It is also a refreshing change from all that calculus-based mathematics. This particular project handout was an attempt to woo those more interested in pure mathematics so it is that side of the subject that is emphasised.

This particular year, two students expressed a desire to do a project involving graphs and networks. One would have thought that this would have been straightforward, however they fell out from the word go although the module leader was not aware of the completeness of the breakdown until three weeks into the project. So unfortunately, by this time it was impossible for either of them to change to another group. The answer to this problem is a pragmatic one: allow both of them to continue as a "group" but also allow them to continue independently minimising the amount of interaction. This is really only possible in a group of two and is particularly easy in a graphs and networks group. One student concentrated on undirected graphs, the other student concentrated on networks. In effect, they both undertook individual mini-projects. By not working together, the students missed the opportunity to cooperate, to interact, to pool resources and partake in general in the benefits of group work. However on the positive side, assessment of the contribution of each was made easy. The module leader was not happy with this solution, but it was felt to be preferable to shoe-horning them together and producing lots of friction and misery but very little else.

The first student was not particularly strong in terms of mathematical ability, so (s)he decided to concentrate on defining various types of graphs and outlining some practical uses of graphs. First of all features such as a complete graph, a subgraph, cycles and paths were defined. The adjacency matrix and the incident matrix were introduced as was the handshaking lemma and its

4. Group Projects

Graphs and Networks

1. <u>Mathematical Nature</u> This project may only involve arithmetic. It can also involve linear algebra and matrices. Reference material: books on graph theory by Ronald Gould, by F. Harary and by Robin J Wilson. A more applications based book is *Mathematics in Communication Theory* by R H Jones and N C Steele published in 1989 by Ellis Horwood. The first part is most relevant for this project, the second more use for Fourier Transforms.

2. <u>Possible Strategies</u> Several different projects are possible based on graphs and networks. Here are two which in themselves have very broad scopes.

(a) Undirected Graphs - look at the properties of trees; binary trees and quad trees and their relationship to the storage of images on software.

(b) Networks - look at the properties of directed graphs; cutsets, fundamental cycles and apply ideas and concepts to the analysis of electrical networks.

3. <u>An Example of a Project Plan</u>

(a) Graph Theory (three people)

    A Study tree searching algorithms

    B Look at computer implementation

    C Look at a practical application such as the storage of images or the parsing of sentences.

(b) Networks (three people)

    A Examine various network flow algorithms e.g. depth first search, FIFO .

    B Investigate more complicated algorithms and their implementation on a computer.

    C Looks at matrix methods for electrical networks and connections to practical applications.

Other topics that could lead to projects are map colouring and tiling.

---

consequences. Here is the handshaking lemma:

In any graph, the sum of all the vertex degrees is equal to twice the number of edges. The proof is very straightforward:

Each edge has exactly two ends, and by the definition of degree it therefore contributes exactly twice to the vertex degree. The result follows.

This gives a flavour of the difficulty (or more strictly the lack of difficulty!) of the mathematics in the opening sections. The discussions of bipartite graphs and unions and complements of graphs was a little more involved, but it is not until applications are addressed that the project really comes alive. This particular student found applications in the social science of relationships between sets of three people with applications to TV soap operas (!), to engineering in the form of the design of waste disposal in a city and the representation of the change of energy use by the population of (the same?) city. Finally there was a solution to a puzzle that was popular in the 1970s called "instant insanity". To give a flavour of this part of the project, this solution is given below.

There are four cubes. The faces of these cubes are either red, blue, yellow or green. The object is to stack the four cubes on top of one another such that when stacked no two adjacent faces have the same colour. Moreover, looking at the four cubes stacked on each other, each vertical face contains each colour once and only once. In order to solve this problem by trial and error, even using a little common sense only reduces the number of possibilities from 331,776 (which is 24^4) to 82,944! Hence the name of the puzzle. First of all, each cube is represented by a graph as shown in Figure 4.1 The graphs are drawn as follows. A cube has three pairs of opposite faces. Draw an edge between vertices if and only if the colours shown (G for green, Y for yellow, etc.) are on opposite faces. Thus each cube corresponds to a graph with four vertices and three edges. The next step is to combine all these graphs into one. This is shown in Figure 4.2. The edges are marked with numbers that tell us from which cube the edge arises. What happens now is that we search for cycles that are subgroups of these graphs. Further, each cycle must have an edge that is labelled with 1, 2, 3 and 4 just once. No repeats and each cycle must have different edge labels between equivalent colours. There are only two distinct cycles of this type and these are given below. This enables us to solve the puzzle as follows: mark each cube 1, 2, 3 and 4 respectively. Use the first cycle H_1 to tell which colours go to the back and front, and the second cycle H_2 to tell which colours go left and right. So on the first cube Y is in front and B is opposite this whereas G is to the right and R to the left. Take the second cube: G is at the front and Y is at the back, R is on the right and B on the left. For the third cube B is at the front with R to the rear, B is on the right and Y on the left. Finally R is at the front, G the back, Y is on the right and G is on the left. The stack so formed solves the puzzle hopefully with the minimum of insanity! The decision has been taken here to give a solution to the puzzle rather than copy verbatim from the project report. Much of the style of the report has been retained, but in the project report itself the explanation is incomplete and in general lacks clarity. It does not mention that the cubes have to be specially designed, otherwise it is possible for there to be no solution at all, or many solutions (not that many!). It all depends on the character of the combined graph. This student lost some marks for the general lack of mathematical depth in the project, but gained marks by showing a breadth of application, a readable writing style and good presentation.

The second student concentrated on matching, assignment and described the Hungarian algorithm for optimising the assignment of tasks to people, the optimisation being in the sense of minimising costs. This student also introduced bipartite graphs, and if considered as a joint project with the first candi-

4. Group Projects

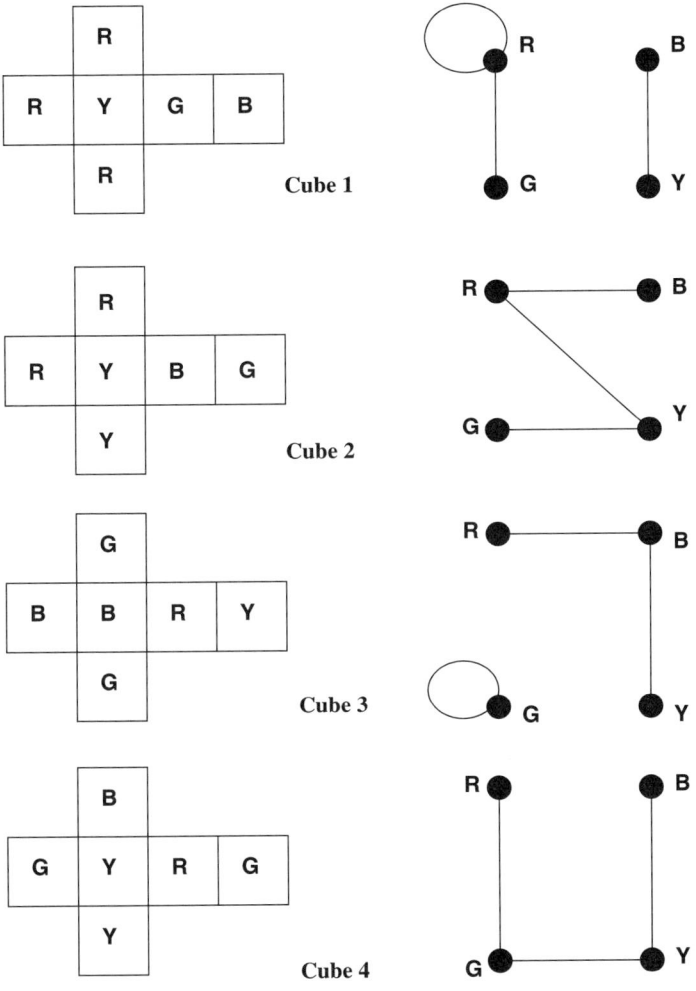

Figure 4.1 The four cubes and their corresponding graph.

date this would be a clear case of overlap. However, they were not penalised for this in this instance as they had had permission to write two distinct reports. Fortunately the overlap was not great and it was blindingly obvious that no copying had taken place as the presentation and notation was so different. Here is an extract from the project that illustrates the kind of matching problem that was tackled.

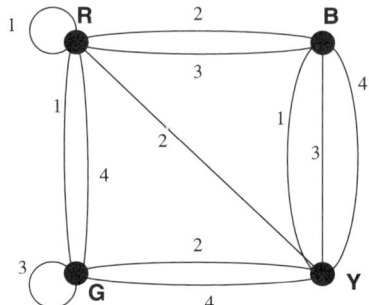

Figure 4.2 The combined graph.

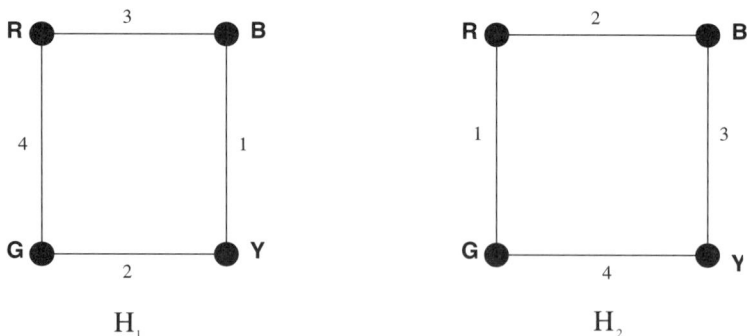

Figure 4.3 The two cyclic subgraphs.

4. Group Projects

A store has four specialist departments, Electrical, Furniture, Food and Hardware. Each department needs someone to manage it. There are four managers, Anne, Barry, Carol and David. Each has varying managerial experience. Anne could manage Electrical or Furniture but not the others. Barry has never worked in the Furniture department but could manage any of the others. Carol has experience managing all departments except Hardware. David could manage Electrical or Hardware. The question is how can the Store Manager best assign the management jobs? The answer is to draw a bipartite graph with A, B, C and D on the left and E, F, G and H on the right. A line is drawn connecting a left vertex (name) with a right vertex (department) corresponding to suitability given the above information. The result is the graph depicted below. In this instance, it is quite easy to match by inspection (or trial and error).

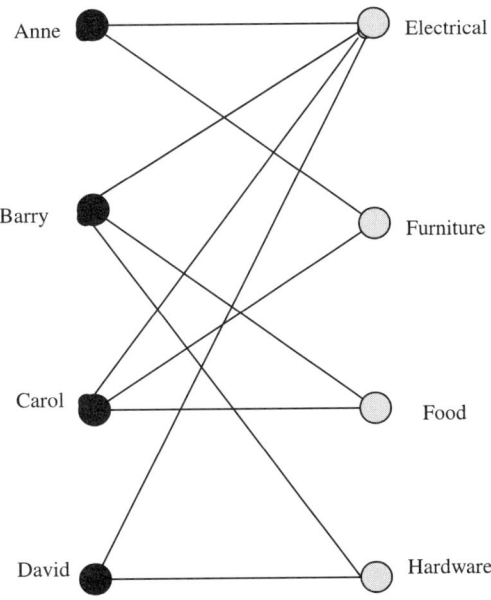

Figure 4.4 The bipartite graph for the matches of manager with department.

For example, one solution is Anne - Electrical, Barry - Food, Carol - Furniture and David - Hardware. There are three other solutions, but the point made by the student a little later is that if there were many more vertices a solution by hand would be impractical. In more complex problems it is no longer a question of merely either being able or unable to do something but there will be varying degrees of suitability. Perhaps each manager has a price (the example needs to be more complex for the analogy to work ideally). In this case a cost matrix can be defined and each edge in the bipartite graph is assigned a value corresponding to the cost of the function on the right being done by the particular individual on the left. An algorithm is required to find a matching that minimises cost. The project then goes on to explain the Hungarian algorithm for doing precisely this. This is quite long and is not repeated here: nevertheless the point is that the student seemed to understand it and he gave an example. What was lacking is any generalisation to more complex problems. There was also no theory to speak of. The project read more like a manual on how to use the Hungarian algorithm. The presentation was good, though not as good as that of the first student. However the level of mathematics was a little higher. This particular student was undertaking computing modules, and his layout and treatment of algorithms betrayed this. Overall his mark was similar to that of the first student.

As a whole, the project was in a sense doomed from the start due to the entire separation of the two students. There were no marks for co-operation between candidates, peer review had to be the same (in percentage terms) for each individual contribution which in this instance was separately marked. It was important that the students were not disadvantaged by the circumstances. However one could argue that their inability to co-operate was a shortcoming that ought to be reflected in the assessment. The problem with this is the apportion of blame. The module leader could not tell whose fault it was that they did not get on, or indeed if it was anybody's fault as such. To deduct marks on hearsay evidence stores up trouble, invites appeals and is simply immoral. The clash between them was not able to be forecast, and since the entire cohort was small there was no solution other than to let them do two individual mini projects. In this sense this project (strictly these two projects) belong in Chapter 3. The lessons here are: first, beware groups of two. Make sure they really want to work together before deciding on the groups. Second, should this situation arise, be pragmatic and put the needs of the students first by changing the rules if necessary. In some respects, make the best of a bad job!

In the next case study, we have a second example of one that did not really work, but for entirely different reasons.

4.5 Fourier Transforms

Fourier series and Laplace transforms form part of a standard mathematics degree course. Fourier series are normally met in year two, either as a stand-alone course or as part of a course on differential equations. Laplace transforms are either a year two or a year three topic, but sometimes they are absent altogether much to the astonishment of electronic engineering students who seem to know about them from day one! Fourier transforms are sometimes met as a limiting process from Fourier series, but typically this is a single lecture and students are not introduced to applications. Fourier transforms are also met within a course on integral transforms, but the same comments apply. This group project led the student towards practical applications but also hinted at more pure mathematical asides. The handout is reproduced below.

Fourier Transforms

1. <u>Mathematical Nature</u>
This is a direct follow on from Fourier series and links to differential equations (ordinary and partial), Laplace transforms and eventually to time series analysis and signal processing.
Reference material is in books e.g. : *Fourier Transform and its Applications* R Bracewell (1986) second edition; and more specialist books such as *Fast Fourier Transforms* E O Brigham (1974). Look under 515.723 in the library, there are lots of books; find one to your taste.

2. <u>Possible Strategies</u>
Start by revising periodic functions and Fourier series. Look at the limiting process by which Fourier transforms are obtained. Note the conditions on the function, they are quite restrictive. Go on to look at specialisms: generalised functions (the impulse function and its relatives); applications to signal processing - the spectrum of a signal, the Wiener–Khinchine relations, Windowing, smoothing, etc. Then particular applications to medicine, environmental science or engineering.

3. <u>An Example of a Project Plan</u>
Suppose that there are five team members and call them A, B, C, D, and E.

A Examines generalised functions, their properties and examples of their use.

B Investigates the theory behind Fourier transforms, the Riemann–Lebesgue lemma, the relationship to linear spaces.

C Researches applications to signals, the Wiener–Khinchine relation (which relates a spectrum to autocorrelation).

D Investigates windows and smoothing (works closely with C).

E Investigates practical applications in electronics, medicine or whatever.

The lecture that accompanied the handout covered the following material. First of all there was a brief resumé of orthogonal function theory, placing Fourier series in context. The Fourier transform was derived as a limiting process. The following points were emphasised: the inverse transform is straightforward, but the Fourier transform of such standard functions as the constant function, polynomials and trigonometric functions do not exist as the improper integral diverges. The class was then reminded about the Dirac-δ function as an example of a generalised function. Finally, the lecture ended by giving the students a description via diagrams of the basic ideas behind the processing of signals.

Armed with this information, a group of three students decided to take on this project. Two of them were friends, but mainly due to both failing so many modules in previous years! They were re-sitting the year. The third member was a mature student, rather distant from the other two both in temperament and work ethic. All three struggled with the finer points of mathematics, although the mature student obviously had a physics/engineering background which was helpful. He also was superb at presentation which proved to be a source of tension between the three students towards the end. Therefore this was always going to be an interesting group. Interesting in the sense of the ancient Chinese saying (curse?) "May you live in interesting times" that is. The scope of the project looked to be fine, and there was a possibility of a good outcome. This possibility was unfortunately not realised.

The mature student took on the role of general manager of the project, and although there were only two others to manage, he had his hands full from the start. He took on the topic of the Laplace transform and its relation to the Fourier transform. After (re)introducing Fourier series he defined the Laplace transform through:

$$\mathcal{L}\{f(t)\} = \int_0^\infty f(t)e^{-pt}dt$$

and derived several properties and elementary results. The impulse function and its use were described before the Fourier transform:

$$\mathcal{F}\{f(t)\} = \int_{-\infty}^\infty f(t)e^{-i\omega t}dt$$

together with its inverse

$$\mathcal{F}^{-1}\{F(\omega)\} = \frac{1}{2\pi}\int_{-\infty}^\infty F(\omega)e^{i\omega t}d\omega$$

were introduced. He then went on to develop the subtleties required by signal processing, notably scaling properties, symmetry and shifting. Finally the discrete version of the Fourier transform (DFT) and the fast Fourier transform (FFT) were described and used in a simple example. This student obviously understood how the FFT is computed. His contribution ended with the introduction of windows and smoothing and a discussion of how these are applied to signal processing. This was a very impressive if isolated piece of work. The second student investigated digital signals, examining the Nyquist frequency, filtering, and concepts of bandwidth and noise. He discussed the difference between amplitude and frequency modulation, and the definition of baud rate.

However his contribution was brief and superficial. In particular one was not really convinced that he knew what he was talking about! The third student concentrated on signal analysis with an account of autocorrelation, spectral density and the connection between them, the Wiener Khinchine relations. Finally he looked at the application to random signals. Again, the same comment could be made. This student's report was altogether unconvincing.

The interim reports indicated that the group did not work together and two were too brief. There was some overlap between the work of the students: however given the track record of two of the students the module leader was pleased to get what he did!

The presentations were acceptable, but once more they were three separate pieces of work with no evidence of co-operation. The mature student did attempt to hold the project together, but with limited success. The work of the other two students was short of what is normally expected from final year students. One was embarrassingly elementary, consisting of little more than school binary arithmetic.

The final report consisted of three separate pieces of work. In the defence of the group, the production was halted due to a catastrophic computer failure, but nevertheless the outcome was disappointing. Apart from the contribution of the team leader which was substantial, the rest was quite elementary mathematically and too short. There was absolutely no attempt at joining the three pieces of work together and the overlaps remained. Clearly there had been very little discussion and two of them were very much done at the last minute. The module leader suspects that the leader was tearing what little remained of his hair out cajoling something out of the other two students before the deadline, although in fairness to the group they all insisted each put in equal effort under the peer assessment process. This is one project where the unbiased input from a dispassionate second assessor would have been particularly useful. Sadly, because this project ran in the first student cohort, such quality systems were not in place and it was not done.

The project as a whole was a pass at the lower second level. Both assessors were aware of the practical difficulties experienced, and allowances were made because the catastrophic computer failure stemmed from a virus that was picked up on a university computer in the open access area! The leader was awarded a first class mark, but the others got lower second class marks. However with hindsight the marks awarded were perhaps over-generous. Their final degree grades were lower in every case.

There was quite a lot to learn from this group project. Perhaps in this case it was not a good idea for these three individuals to operate as a group. The leader was much older and wiser than the others, but unfortunately he also was not the best communicator in the world. Although his personal contributions were excellent in quality, well presented and timely, this tended perhaps to intimidate already struggling students who felt that they could never match this performance. The leader was also not the best at encouraging and bringing on his younger colleagues, not least because he was himself finding some of the more advanced mathematics difficult. That he did so well was entirely due to sheer hard graft not because of innate mathematical ability. He certainly should have opted for an individual project. One might say this is indeed what he did! The other two got through, just. As their performance in this module was not out of line with their performances elsewhere, the outcome here could be said to be as expected.

4.6 Orbital Motion

It is sad to reflect, speaking as an applied mathematician, that it is quite common these days for there not to be much in the way of compulsory mechanics in many mathematics degrees. The positive side to this is that there are many mechanics based topics which are familiar to the lecturer but outside present day syllabuses that can provide rich source material for projects. The specification for a project based on mechanics has therefore to be very broad. The decision was made to try and encourage those whose mathematical preferences lay more towards pure mathematics to dip their toe into applied mathematics. To do this, the emphasis of the handout was on the Lagrangian description of mechanics. It is reproduced here.

To accompany this handout, the lecture began by discussing Newton's laws of motion, particularly the second law. Configuration space was defined and the notion of degrees of freedom was explored. The concepts of potential energy and kinetic energy were outlined and students were reminded of the conservation of energy. The Lagrangian was defined and calculated for simple systems such as the pendulum and motion under gravity with which the students were (should have been) familiar. The Hamiltonian and general principles such as the Principle of Least Action were discussed. The quadratic form was revised, and students were shown how the Lagrangian of a linear vibrating system avoided many of the difficulties of the traditional method and was amenable to calculations involving the matrix algebra associated with eigenvalues and eigenvectors. Finally the Poisson bracket and its extension to quantum mechanics was mentioned, though no details were given. Emphasis throughout was on the very general application of Lagrangian mechanics.

Virtually the entire class was seduced by this description, and in order to avoid disappointment, *four* parallel projects involving mechanics ran in this

---

Analytical Dynamics

1. <u>Mathematical Nature</u>
This project has the potential to be entirely analytical and can be viewed as an application of linear algebra (matrices, eigenvalues etc.)
Reference material is *Classical Mechanics* by T W B Kibble and F H Berkshire (1996)(4th Edition), *Introduction to Analytical Dynamics* by N M J Woodhouse (1987) and *Classical Mechanics* by H Goldstein (1980) (2nd Edition). The code 531 in the library should reveal many others and it is worth a browse either on foot or electronically.

2. <u>Possible Strategies</u>
Start from the concept of a dynamical system whereby a system is described by the position and velocities of its constituent particles. This is a function of n variables. The general field of analytical dynamics is about determining equations obeyed by these functions. It can be quite "pure mathematical" and only relate to Newtonian mechanics when specific applications are tackled. Historically it arises from the work of Joseph Louis Lagrange in the late 18th century. French mathematics around the time of the revolution was very active, very good and by and large pure mathematics.

3. <u>An example of a project plan</u>
Five team members A, B, C, D and E.

A Researches canonical variables and Lagrange's Equations

B Researches Hamilton's Equations and the Principle of Least Action

C Researches modern dynamical systems in the context of analytical dynamics

D Relates analytical dynamics to Lie Algebras, Astronomy, Relativity or Quantum Theory

E Solves some simple problems using Lagrange's Equations.

---

particular year. Obviously they had to be quite distinct. One was on the analysis of a spinning top and associated gyroscope. One was on deriving Hamilton's equations and how these are extended to include relativistic effects. Two of them did not in fact involve very much in the way of analytical dynamics but were traditional mechanics. One involved the application of dynamics to astronomy, and only touched on a Lagrangian description near the very end, and the fourth project focused on orbits and their stability. This last project is the one we shall describe in more detail here. The fact that it deviated very much from the handout is not a problem. The object of the handout is to give the student or group of students ideas, and if they drift away from the given description all is well provided the lecturer approves the final plan. In past times, a project based on orbits would have overlapped considerably with second year, perhaps even first year, mechanics. This is not so obviously the case these days, but there will be some traditional mathematics degrees in which the overlap will still be there. In Plymouth there is a little on orbital motion in the second stage of the degree, but not a significant amount. Certainly not enough to trouble those concerned with projects duplicating taught material. Of course, even if the course has a traditional mechanics based module with much on orbits, the possibility of pursuing projects based on orbits is still there but has to reflect the more advanced student knowledge. For example, something on the stability of satellites perhaps involving numerical techniques would always be a possible project. Another option would be to look at relativistic effects, exploring the motion of Mercury. The students who tackled this particular project had a little previous knowledge, but decided on a project scope that covered a broad range within the general area of orbits although there was an attempt at an analysis of the stability of orbits. The three students (all female as it happens) really did work together writing at least a chapter each, but combining over others. The interim reports were all very well done, which was not surprising given their academic records. The verbal presentation is best described as courageous. It could have been brilliant, but unfortunately they lacked the bravado and it was not entirely the right vehicle. What they did was to present a mock interview where one student questioned one other whilst the third used the overhead projector. The "expert" was by common consent the most able academically and obviously understood the subject thoroughly, but the style of the presentation was not geared to bring out the knowledge of the other two. The questions could have been better thought out. The advice of the lecturer was that this was a novel idea that did not really succeed. It would have been more at home in an Open University summer school where the emphasis is perhaps more on enjoyment and group participation!

The final project report was well done. It opened with a chapter on history and astronomy which was scholarly. This finished with Kepler and his laws. In chapter 2 the polar co-ordinate system was introduced and the acceleration

$$\left(m\{\ddot{r} - r\dot{\theta}^2\}, \frac{m}{r}\frac{d}{dt}(r^2\dot{\theta})\right)$$

derived. Then followed two chapters on the geometry of conics including detailed descriptions of the ellipse. Then Kepler's three laws were proved. This included the proof that orbits were ellipses using reciprocal polar co-ordinates. There were some unfortunate layout problems at this point. The chapter headings became inconsistent, the page numbering disappeared and the numbering of equations was confused. This was a consequence of lack of time and some printer problems towards the end of the project. The final chapter concerned the stability of orbits and took the project to another level of difficulty. Methods of expansion were used and the motion of the disturbed body compared to the simple harmonic oscillator in order to find periods. In the project, the (radial) equation of motion of an orbit in the form

$$m\ddot{r} - \frac{mh^2}{r^3} = f(r)$$

was expanded using $r = x + a$, and the equation

$$m\ddot{x} - \left(\frac{3}{a}f(a) + f'(a)\right)x = 0$$

analysed for stability using a general power law. Overall the students were a little disappointed with the final project, mainly due to the production problems leading to the style and layout glitches. The assessors read the project and were in general pleased with its mathematical content. However, the mathematical standard of the work on the stability of orbits was some way above the rest. Care had to be taken not to use this as a negative influence, condemning the rest as not worthy of a similar grade. The problems with the layout were enough to bring the overall mark down from first to upper second. The message here, given time and time again, and ignored by students everywhere is not to leave things to the last minute. In their finals, one student got a first and the other two awarded upper seconds and in every case the group project mark was their worst, so maybe their disappointment was in some way justified!

4.7 Conclusion

This chapter is an account of the experience of the author in running group projects in mathematics. In some senses, a group project in mathematics is much the same as a group project in another discipline, say engineering. However there are distinctive features. In engineering, the partitioning of tasks needs to be much more carefully done and the whole more managed. There is usually something to be made involving technicians who have other things to do and their own agenda, and there is usually an end product that should be tested. The engineering group project is more multi-disciplinary and aspects such as health and safety, and economics (cost) rear their ugly heads. The author was, in a previous existence, responsible for inter-disciplinary group projects in engineering so these are words written from experience. What the mathematics group project gives is the opportunity for the student to work with fellow mathematics students. This helps them to be aware of the strengths and weaknesses in others and themselves, to tolerate both and to work towards a common goal. It also gets the students introduced to managerial skills such as conducting themselves in meetings perhaps even chairing them, managing their time, keeping to deadlines set by themselves, and listening to others. With few exceptions, students who have participated in a group project have found it a very worthwhile experience.

In the next chapter, case studies are explored. These are in a sense a half way house between a group project and a standard lecture course. We finish this chapter with a section on further suggestions for group project topics.

4.8 Further Suggestions

Normally, at this stage there would be a number of exercises on the contents of the chapter, but as in the last chapter that does not work here. Instead there follow a few suggestions for group projects. Some of them have been tried, but some await trial.

1. **Financial Mathematics**

    With the Thatcher era of the 1980s, there was a surge in interest in business and finance. In particular, a number of research mathematicians became interested and successful in applying mathematical modelling to the behaviour of the stock market. In particular, a number of successes were attributed to the Black–Scholes equation:

    $$\frac{\partial f}{\partial t} + rS\frac{\partial f}{\partial S} + \frac{1}{2}\sigma^2 S^2 \frac{\partial^2 f}{\partial S^2} = rf$$

    which describes the behaviour of the price f given information about stocks S. The rest of the notation is r is the interest rate and σ is a measure of the volatility of the stock and is related to the standard deviation of the proportional change in stock price in an increment of time. There is a great deal of scope here for group projects. All the students need to have an interest in and understand finance and investment. A group of five can then look at the following tasks. One compares and contrasts discrete and continuous models of finance. This student could also look at the history of financial modelling. One student could investigate Ito calculus which is a calculus developed by Kiyosi Ito (1915–) to describe stochastic variation. He developed it for biological and physical systems, but it is well suited to model finance. This student could use Ito calculus to derive the Black–Scholes equation, but this does require a background in probability and the analysis of time series. A more direct approach based on seven assumptions is probably a better alternative. The third student could solve this differential equation in a variety of cases. It transforms into the standard diffusion equation, but the boundary conditions are troublesome. Laplace transforms can also be used, but inversion is hard and the solution usually only obtained in terms of an integral. The fourth student can examine numerical methods of solving the Black–Scholes equation. This can involve setting up a semi-implicit Crank–Nicolson scheme and writing a program either in

FORTRAN or a language such as MAPLE. Finally the fifth student should look at the shortcomings of the Black–Scholes model and investigate alternative models. It is certainly true that the Black–Scholes model (sometimes called the Black–Scholes–Merton model) does not seem to work now and is in need of modification. This is a clear case of the application of a model influencing the behaviour of the system itself. Perhaps students could think along the lines of Heisenberg's uncertainty principle!

2. **Normal Modes**

The whole concept of normal modes or natural frequencies/resonance makes a good topic on which to centre a group project. Presumably, most students would have met the concept of resonance when discussing the solution of the equation for simple harmonic motion. Normal modes might have been met in a particle mechanics course or more theoretically in a linear algebra module in connection with the application of eigenvalues and eigenvectors. So the idea that a linear system possesses natural frequencies that can be excited by forcing at these frequencies should not be entirely new. The introductory talk can be adjusted to cater for the particular needs of the student body and to get them up to an appropriate level. Let us look at some student tasks that can be assigned within this general heading.

One sub-project could be to examine the solution of the Helmholz wave equation
$$(\nabla^2 + \lambda^2)\phi = 0$$
where $\lambda = \omega/c$, ω = frequency and c = celerity (wave speed). If this is solved in one dimension then there is one natural frequency λc which is the single "normal mode". If solved in two or three dimensions under a cartesian co-ordinate system, then there are nodal lines or planes corresponding to a standing wave solution when the whole vibrates in a finite domain. From a mathematical point of view the solution to the Helmholz wave equation in cylindrical geometry is of interest as it yields Bessel's differential equation. In spherical geometry a variables separable assumption results in Legendre's equation. In both of these cases, there are a wealth of results that can be derived and which take the student through some of the properties of these special functions. Particularly important are the orthogonality relations which in turn lead to Fourier–Bessel series in the one case and spherical harmonics in the other. The graphical capability of the computer can be used to good effect to visualise these oscillations, and if practical applications are desired there are plenty. The vibration of a membrane, the propagation of electromagnetic waves, and geophysical oscillations are all worth exploring.

On a different tack there is much mileage in exploring the difference between linear and non-linear systems. Linear systems vibrate at natural frequencies that are independent of the initial conditions that get them going. On the other hand non-linear vibrations oscillate at frequencies that are dictated by initial conditions. As an example, one could use van der Pol's equation:
$$\ddot{x} + e(x^2 - 1)\dot{x} + x = 0$$
where e is a constant (which could be assigned the value $\frac{1}{4}$ say) and explore the concept of a limit cycle. Again, one must be lead here by the taught modules experienced by the student to avoid duplication.

Another avenue to explore might be the connection between normal modes and Green's functions. If Green's functions are nowhere to be seen on the course, then an elementary treatment might be expected whereby the Green's function of a system is defined as its response to an impulse, i.e. the system's response to being struck. More advanced approaches will involve two or three dimensions and solutions in various domain shapes.

Finally there is the role of damping in reducing the amplitude of oscillations at natural frequencies or normal modes. The student would have already met mass-spring-damper systems, in which case more complex systems need to be studied. This subject could involve examination of seismograph records to determine damping coefficients.

3. **Water Waves**

   Some mathematics courses have modules devoted to waves in general or to specific kinds of waves such as waves on water, electromagnetic waves or waves on membranes. This possibility is mentioned in connection with the case study on traffic waves in the next chapter, but only a few specialist mathematics courses will have anything except one dimensional waves on strings. Water waves will then provide an opportunity to link the development of wave theory with fluid mechanics.

   One student will certainly have to concentrate on linearising the fluid mechanics, including the dynamic boundary condition (pressure is a constant on the free surface) and deriving that one does indeed need to solve the wave equation:
   $$\nabla^2 \phi = \frac{1}{c^2} \frac{\partial^2 \phi}{\partial t^2}$$
   even if it is restricted to one dimension:
   $$\frac{\partial^2 \phi}{\partial x^2} = \frac{1}{c^2} \frac{\partial^2 \phi}{\partial t^2}.$$

In addition, there could be an investigation of the perturbation technique for solving the non-linear problem that has been used so successfully over the years by Civil Engineers to design codes of practice.

This leads to a second exploration that moves squarely into the realm of Civil Engineering, that is the calculation of wave forces. The use of various drag laws, the approximations and assumptions made are all important aspects for the student to understand. Wave forces on fixed and floating structures, how such forces vary with depth as well as the impact of waves on the shore are all possible topics here. The use of design loads and the use of engineering safety factors gives a student mathematician useful insight into the real world of the engineer. It is also possible to extend this investigation into the calculation of wave forces due to a spectrum of waves, although a great deal of guidance is usually required. Engineers do tend to oversimplify such problems.

Another student could examine wave ray theory and its role in deriving results in water waves. Non-linearity and caustics can be investigated here too. On a more practical level, using wave ray theory to investigate the approach of a wave train to a beach or in the design of harbours and marinas also leads to fruitful mathematics. Depending again on their background, the student could tie this up with the theory of optics.

Finally, there is the application of water wave theory to open channel flow and the whole area of river hydraulics. This is a vast topic and usually constitutes a large part of hydraulics in a Civil Engineering degree. Useful areas for a mathematics student to concentrate on are the role of waves in moving sediment, the interaction between waves and river flow (bringing in the Froude Number and stability criteria) and finite and variable depth. Allowing the depth to vary in some idealised way can lead to special functions and other interesting mathematics.

4. **Numerical Solutions to Differential Equations**

These days, all mathematics courses introduce students to the numerical solution of ordinary differential equations. This usually takes place in the second year where the syllabus covers Euler's method, modifications thereof and Runge–Kutta methods. Formal examination questions on this topic are limited, so there is usually a coursework involving exploring the solution to some problem which calls (using FORTRAN or MAPLE perhaps) a routine for solving ordinary differential equations. It is usual for the student to have written a simple program to solve a boundary value problem in one spatial variable, but for complex situations commercial software (NAG routines) are called. Again, different universities will have varying amounts of this type of material, so the initial introductory talk has to start from a carefully set level. In the unlikely event of there being no numerical solutions to ODEs anywhere on the course, then all students can start with the same clean sheet. Here, let us assume all students are familiar with Euler's method and Runge–Kutta methods.

One student could explore the different truncation errors of the different methods. Using Euler's method as a reference, the second and perhaps fourth order Runge–Kutta methods could have been covered in class. In which case the student needs to calculate the error of more sophisticated methods such as the multi-step Adams–Bashforth and Milne methods.

Another possibility is to see how the numerical methods developed for ODEs generalise to certain types of PDE. The first step is to tackle second order ODEs if this has not yet been done. After this, there could be a need to restrict attention to explicit methods in time or elliptic equations which do not involve time. The calculation of errors is not all that different between ODEs and PDEs, but there is plenty of scope for projects to explore different methods of computation. There is also the opportunity for the student to produce visual aids for understanding the methods of solving two dimensional or even three dimensional problems. If the emphasis is on studying different methods it might be best to restrict the study to one equation, say Laplace's equation

$$\nabla^2 \phi = 0.$$

On the other hand, the student could tackle a real problem and solve it numerically. A laboratory set-up such as a floating cylinder is reasonably straightforward: a simplified storm surge prediction based on shallow water of constant depth is harder, but not impossible.

Given that numerical methods for solving ODEs and PDEs usually end up with having to invert large matrices which may or may not be sparse, a student could look at the problem of large matrix inversion by numerical means. This could include the development of sparse matrix solvers, use of complex arithmetic and the development of user friendly software.

5. **Tops and Gyroscopes**

As was mentioned in Section 4.6, there is often very little compulsory mechanics in a mathematics degree course these days. Often the amount covered only equates to what used to be in 'A' levels thirty odd years ago. Therefore a third year mathematics student typically has sparse knowledge of Newton's laws as applied to particles, including perhaps orbital motion (but see Section 4.6). He or she also knows about moments of inertia and can calculate them as examples of multi-variable calculus but typically cannot analyse rolling bodies. Given the restricted nature of the topic, this is only really suitable for a group of five students if there is very little mechanics in the rest of the course. Let us assume all students know about particle mechanics, and understand the concept of moment of inertia but little else.

The first important step to take is for *all* the students to be familiar with Euler's dynamical equations. This is an instance where the entire group needs to get to grips with a new piece of work and where they need to elect a member who can write it up and convey the result to the rest of the group in an understandable way. The first student undertakes to derive Euler's equations and perhaps goes on to solve these equations in the special case of a spinning axisymmetric top. Precession is important, nutation less so.

One student can look at the life and times of Lagrange and how analytical dynamics was developed as a consequence of applying mathematically rigorous thinking to Newton's laws. This student would also need to develop rigid body mechanics of the type that would lead up to Euler's equations, that is two dimensional rolling bodies.

A third student could look at applications such as the principle of the gyroscope and the development of the gyrocompass. Some practical problems would then be solved.

There is potential for this project to go on to feature Lagrangian and Hamiltonian dynamics, especially if quantum mechanics features as a taught module where the Poisson bracket is important. In general it is better to link group projects to work done by all the students. This is beneficial and a project that introduces the Hamiltonian and the concept of action can be very useful in the study of quantum mechanics.

6. **Boolean Algebra**

Boolean algebra has an interesting place in mathematics. The underlying concept in the form of the algebra of binary numbers is met at school. However, sometimes there is no Boolean algebra as such in any of the three years of a mathematics degree course! Most students meet abstract algebra (groups, rings and fields) and Boolean algebra can be introduced as an example. However, it has such a distinctive feel that this group project can virtually be started from scratch.

One student can look at the definition of union and intersection, the basic laws of associativity, distributivity and commutativity, complementarity and the De Morgan laws.

One student can look at the application to switching circuits and the wealth of applications in electronics. This nicely leads on to NAND and NOR gates and the use of Boolean algebra to simplify circuits. How to design a circuit to control the central lights in a room with three doors is a favourite. Designing lift buttons so that the nearest one out of three or four comes is another.

The applications of Boolean algebra to propositional logic and the design of expert systems is another fruitful area for a student. This could also include an analysis of the various methods of proof which by this stage should be familiar to the student.

There will be a certain amount of interplay between this group project and any option on control theory that might be running. Again it is up to the lecturer to keep an eye on potential overlaps.

7. **Archimedes and Apollonius**

This group project cannot run if the history of the mathematics of Archimedes and Apollonius forms part of a module. These two encompass the development of mathematics (principally geometry) in ancient Greece, and the subject area is so vast that only a few hints of possible sub-projects are given.

Here are a few investigations linked with Archimedes:

the Principle of Archimedes (hydrostatics and the famous "Eureka!" moment);

the method of indivisibles and his calculation of areas, predating the calculus by 2000 years;

the Archimedian solids;

his work on cycloids.

for Apollonius we have:

his work on ratios;

the conic sections;

various astronomical calculations including cycles and epicycles, and those involving comparing the dodecahedron and the icosahedron.

There are many other possibilities. In general the History of Mathematics can provide a good vehicle for both group and individual projects, but there is a health warning. Make sure that each student engages with the mathematics and not just with the person and the culture of the time. As important as these latter are, this is a *mathematics* group project.

5
Case Studies

5.1 Introduction

If the idea of plunging straight into student centred projects seems too much of a leap into the dark, an alternative is to provide case studies. This chapter shows how to do this largely through examples. The traditional lecture course (or "module" in these days of unitised courses) consists of a set of aims with the expected learning outcomes together with a syllabus. The student attends lectures, and there are usually (but increasingly less frequently) problem classes where the student can receive help with prescribed problems set by the lecturer. The whole is wrapped up by an end of module examination. For most modules there is in addition a coursework element whereby each student independently is assessed on a longer piece of work which has to be completed over about three weeks or a month, but this rarely exceeds 25% of the total assessment load. Commonly, one piece of coursework is set to be done two thirds of the way through the module. The exception to all this is numerically based mathematics which lends itself more naturally to coursework based assessment rather than the formal end of unit examination. Modules involving largely numerical work can be entirely coursework assessed, but 50% is more common. Of course, the statistics parts of a mathematics degree are peppered with potential case study material but these are outside the scope of the present text.

The case study approach is in a sense a half-way house between a formal lecture course and a group project. The lecturer introduces a scenario, usually taking two or three lecture hours to do so. After this presentation, each student working in isolation or in groups is assessed on their understanding of the study. There is no extra research by the student, but the student is expected to solve some extra follow-up material not presented by the lecturer.

Assessment is usually via coursework but there could be an examination too. There are some further words on this in the chapter summary at the end. Some suggestions for assessment tasks are given for each of the case studies. These take the form of alternative analytical forms, alternative assumptions, considering one feature and exploring further, etc. Here is the first of six examples of case studies. The form this chapter takes therefore is a sequence of quite detailed descriptions of case study examples. These are designed mainly for the lecturer so that they can provide a vehicle for a class based case study, but they will also be useful for students undertaking the studies. It could be the opinion of some that the description given is a little overfull and lands the lecturer with the problem of not leaving enough for the student to do. This author would rather err in this direction than not giving enough detail. Besides, all the examples are open ended and it is always possible for any lecturer worth his or her salt to extend the problems. Let us introduce the first example now.

5.2 Ocean Surface Dynamics

This case study is unashamedly based around the research interests of the author. It concerns the way the ocean responds when a wind blows over it. It requires students to be familiar with vector calculus, differential equations and, to some extent, fluid mechanics, although the amount of fluid mechanics required as a prerequisite can be minimal and catered for by the style of delivery. The most difficult aspects for students to follow are not the technical facility with any particular mathematics, but with what can be neglected and how to model what is retained. There are textbooks about how to model the movement of the ocean, most of them intended for postgraduate researchers which are too sophisticated for even final year students, so the case study has to be presented carefully.

The first place to start this case study is with a description of the accelerations and forces that act on the ocean. The atmosphere and ocean are fluids attached to a rotating earth by gravity. There are three accelerations: one due to the accelerating fluid which is called the particle acceleration, the second due to the rotation of the earth called the Coriolis acceleration and a third due to any fluid particle not being isolated (it is surrounded by other particles in much the same way as a rigid body although it is not rigid of course): this is the advective acceleration. The first of these is the rate of change of velocity and is neither a difficult nor a new concept. The other two are trickier. The third acceleration is not important as an acceleration, but provides the source for the turbulence model to follow later. We return to it then. The second is important and the following mathematics describes how it is included in the equations of motion. First of all Figure 5.1 shows the earth and its rotation about the North–South axis. An arbitrary point on the earth's surface is la-

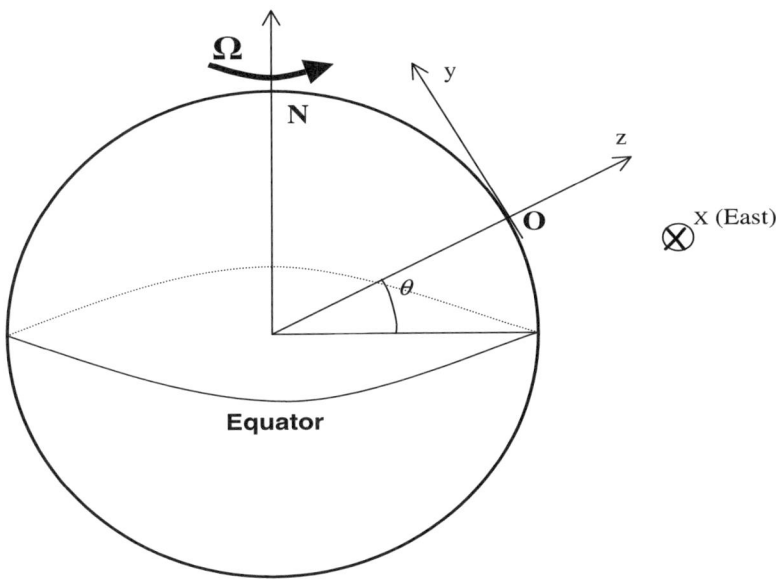

Figure 5.1   Motion relative to a rotating earth.

belled O and is the origin of local (x, y, z) co-ordinates. x points east, y points north and z points up. In this co-ordinate system shown in Figure 5.1 the velocity of the fluid is **u**, but of course the origin is moving with respect to the centre of the earth. (The earth's centre will be assumed fixed as its motion around the sun has such a large radius of curvature that it is straight to a very good approximation.) The motion of the point O is by virtue of the angular velocity of the earth. In the plane of the line of latitude that contains O, the x and y axes rotate with angular velocity $\Omega \sin\theta$ where θ is latitude. This is the vertical (z) component of Ω. The vector Ω has direction south pole to north pole. Thus, when the derivative (rate of change) with respect to time of any vector is calculated, and this calculation is performed relative to the rotating axes, the changes of the unit vectors that point along Ox and Oy, call them $\mathbf{i}$ and $\mathbf{j}$, need to be taken into account. So, take any vector $\mathbf{A}(t)$ which is a function of time. Then the calculation of its rate of change proceeds as follows:

$$\frac{d\mathbf{A}(t)}{dt} = \frac{d(A_1\mathbf{i} + A_2\mathbf{j})}{dt} = \frac{dA_1}{dt}\mathbf{i} + \frac{dA_2}{dt}\mathbf{j} + A_1\frac{d\mathbf{i}}{dt} + A_2\frac{d\mathbf{j}}{dt},$$

where A_1 and A_2 are the components of $\mathbf{A}(t)$ in the x and y directions respectively. By examining how the axes change in an infinitesimally small time, it can easily be shown that

$$\frac{d\mathbf{i}}{dt} = \Omega\mathbf{j}, \quad \text{and} \quad \frac{d\mathbf{j}}{dt} = -\Omega\mathbf{i}$$

and hence

$$\frac{d\mathbf{A}(t)}{dt} = \frac{dA_1}{dt}\mathbf{i} + \frac{dA_2}{dt}\mathbf{j} + \Omega A_1\mathbf{j} - \Omega A_2\mathbf{i}.$$

The first two terms are the rate of change of $\mathbf{A}(t)$ with respect to time as if the axes were not rotating, so the extra two terms represent the effects of the rotation. These can be put succinctly in terms of vector quantities by using the cross product. Thus we can write:

$$\left[\frac{d\mathbf{A}(t)}{dt}\right]_{fixed} = \left[\frac{d\mathbf{A}(t)}{dt}\right]_{rotating} + \Omega \times \mathbf{A}(t).$$

The derivation of this has been two dimensional and may not look very general. However, it is mathematically rigorous as the above vector equation is co-ordinate independent, and a mathematical result that is co-ordinate independent but derived using specific co-ordinates (in our case two dimensional Cartesian co-ordinates) remains true in all orthogonal curvilinear co-ordinate systems. This is a useful theorem and illustrates the usefulness of pure mathematics (as if that were necessary!).

The Coriolis acceleration is now easily derived by a double application of the formula. If $\mathbf{r}$ denotes the position vector relative to an inertial frame of reference (the centre of the earth for example) then we have

$$\frac{d^2\mathbf{r}}{dt^2} = \frac{d}{dt}\frac{d\mathbf{r}}{dt}$$

where all derivatives are relative to fixed axes. The left-hand side is the true acceleration. In terms of actually measurable quantities which are of course

5. Case Studies

measured relative to a rotating frame of reference we thus have:

$$\frac{d}{dt}\frac{d\mathbf{r}(t)}{dt} = \left[\frac{d}{dt} + \mathbf{\Omega}\times\right]\left[\frac{d}{dt} + \mathbf{\Omega}\times\right]_{rotating} \mathbf{r}(t)$$

$$= \left[\frac{d^2\mathbf{r}(t)}{dt^2}\right]_{rotating} + 2\mathbf{\Omega}\times\left[\frac{d\mathbf{r}(t)}{dt}\right]_{rotating} + \mathbf{\Omega}\times(\mathbf{\Omega}\times\mathbf{r}(t)).$$

The Coriolis acceleration is the term

$$2\mathbf{\Omega}\times\left[\frac{d\mathbf{r}(t)}{dt}\right]_{rotating}$$

The term

$$\mathbf{\Omega}\times(\mathbf{\Omega}\times\mathbf{r}(t))$$

is the centripetal acceleration which is directed towards the axis of rotation of the earth but is so small compared to gravity that it is usually ignored. We have also assumed that the angular velocity of the earth $\mathbf{\Omega}$ does not vary with time. (Otherwise there would be yet another term to consider.)

With respect to the chosen co-ordinate system (see Figure 5.1) the Coriolis acceleration is

$$2\mathbf{\Omega}\times\mathbf{u} = 2(0, \Omega\cos\theta, \Omega\sin\theta)\times(u, v, w)$$

where $(u, v, w) = \mathbf{u}$ is the fluid velocity. The double use of the symbol u is less confusing than alternatives. u is the easterly current, v the northerly current and w the upward current. This latter is of course very small and is neglected except for very specialist modelling. w is always neglected when compared with u or v, and it is this approximation that gives the Coriolis acceleration the components $(-fv, fu, 0)$ where $f = 2\Omega\sin\theta$. f is called the Coriolis parameter, it is twice the vertical component of the earth's angular velocity, and it is only this component that plays an important part in the dynamics of the ocean and atmosphere. In a mathematics course it is usual to give the details. Expanding the vector cross product we get

$$2\mathbf{\Omega}\times\mathbf{u} = 2\begin{vmatrix} \mathbf{i} & \mathbf{j} & \mathbf{k} \\ 0 & \Omega\cos\theta & \Omega\sin\theta \\ u & v & w \end{vmatrix}$$

$$= 2(\Omega w\cos\theta - \Omega v\sin\theta, \Omega u\sin\theta, -\Omega u\cos\theta).$$

As $w \ll v$ the $\Omega w\cos\theta$ term in the first component may be safely neglected. Moreover, vertically the balance is between gravity and the vertical pressure gradient. As these both have a magnitude of about 10^3 times $\Omega u\cos\theta$ this latter term too can be cast aside. This vertical balance is called hydrostatic pressure and expresses that the pressure at a point in the sea is due to the weight of water above it rather than any dynamic effects, be they Coriolis or anything else. The result is that only the vertical component of the earth's rotation enters the dynamics and as we stated above the Coriolis acceleration can be expressed as $(-fv, fu, 0)$ to a good degree of accuracy, where $f = 2\Omega \sin\theta$.

In order to see what forces act on the ocean we consider an infinitesimally small cube of sea. Gravity acts to pull this downwards and there are pressure forces acting that give forces across opposite faces of the cube proportional to the gradient of the pressure. Both of these are standard forces which students would have met in a course on fluid mechanics. If this is new to the students, then elementary vector calculus of the type

$$-\int_V \rho g \mathbf{k} dV + \int_S (-p\mathbf{n}) dS = -\int_V \rho g \mathbf{k} dV - \int_V \boldsymbol{\nabla} p \, dV = \int_V (-\rho g \mathbf{k} - \boldsymbol{\nabla} p) dV$$

needs to take place. (**k** is the unit vector in the vertical direction, and **n** is a unit outwardly drawn normal to the infinitesimal cube.) The final force to consider is due to friction. The representation of stress in a fluid will be new to most students, so some detail is required. However we first return to the advective acceleration. For those familiar with fluids, this manifests itself in the difference between standard time rate of change and rate of change following the fluid. The first is termed the Eulerian view, the second the Lagrangian view. Mathematics students are often happy with the following explanation in terms of calculus.

Consider a function of four variables, x, y, z, t say $F(x, y, z, t)$. Straightforwardly it has a time derivative

$$\frac{\partial F}{\partial t}.$$

This expresses the way F changes explicitly with time t. However, suppose that x, y and z all depend on t too. The total derivative of F with respect to t would then be expressed via the chain rule as

$$\frac{dF}{dt} = \frac{\partial F}{\partial x}\frac{dx}{dt} + \frac{\partial F}{\partial y}\frac{dy}{dt} + \frac{\partial F}{\partial z}\frac{dz}{dt} + \frac{\partial F}{\partial t}$$

or

$$\frac{dF}{dt} = (\mathbf{u}.\nabla)F + \frac{\partial F}{\partial t}.$$

Here we have used that

$$\mathbf{u} = \left(\frac{dx}{dt}, \frac{dy}{dt}, \frac{dz}{dt}\right).$$

Putting the arbitrary F equal to the velocity **u** now gives the result

$$\frac{d\mathbf{u}}{dt} = \frac{\partial \mathbf{u}}{\partial t} + (\mathbf{u}.\nabla)\mathbf{u}.$$

The first term on the right hand side is the Eulerian rate of change often expressed as the rate of change observed from a fixed point. The left hand side is the total or Lagrangian rate of change often expressed as the rate of change following the fluid. The second term on the right is the difference, which in some contexts (water waves) and after some correct time averaging is called the Stokes' drift. It is now useful to write down the vector equation of motion for the ocean. It is the fluid equivalent of Newton's second law, called the Navier–Stokes' equation, for a rotating fluid.

$$\frac{\partial \mathbf{u}}{\partial t} + (\mathbf{u}.\nabla)\mathbf{u} + 2\mathbf{\Omega} \times \mathbf{u} = -\frac{1}{\rho}\nabla p + \mathbf{g} + \text{friction}.$$

Let us now consider the form that friction takes. We shall then be ready to look at this equation in component form.

The "friction" being discussed here is not molecular friction of the type met in mechanics when objects slide or do not slide down planes, but turbulence. As there are learned books devoted to modelling fluid turbulence (e.g. the classical text by A. A. Townsend (1954)) only a brief introduction is given here. The basic idea in applying turbulence ideas to modelling the sea is that any current in the sea can be thought of as a mean flow that may be time dependent but is steady on a short time scale of a few seconds, plus a small random deviation. Mathematically $\mathbf{u} = \bar{\mathbf{u}} + \mathbf{u}'$ where $\overline{\mathbf{u}'} = \mathbf{0}$ and the overbar denotes the mean taken over a few seconds. This is now inserted in all the governing equations and the overall time average (again over a few seconds) taken. All is as expected, except for the non-linear advective acceleration which produces the extra non-zero term

$$\overline{(\mathbf{u}'.\nabla)\mathbf{u}'}$$

on the left hand side of the Navier–Stokes' equations. The small random deviation $\mathbf{u}'$ is of course unknown, but fortunately the non linear non-zero term is the only place in the equations where it occurs. It is called the Reynolds stress term and although it may look harmless, it actually contains 9 terms consisting of all pairs of combinations of the components of $\mathbf{u}' = (u', v', w')$ with itself. One of the simplest assumptions to make is to relate appropriate parts of this Reynolds stress to gradients in the mean flow. The Reynolds stress is in fact a tensor (strictly a second order tensor) which is made up from the nine components and can be represented by a 3×3 matrix. It can be succinctly written τ_{ij} where the suffices i and j independently run from 1 to 3. In detail

$$\begin{pmatrix} \tau_{11} & \tau_{12} & \tau_{13} \\ \tau_{21} & \tau_{22} & \tau_{23} \\ \tau_{31} & \tau_{32} & \tau_{33} \end{pmatrix} = \begin{pmatrix} -\rho\overline{u'^2} & -\rho\overline{u'v'} & -\rho\overline{u'w'} \\ -\rho\overline{v'u'} & -\rho\overline{v'^2} & -\rho\overline{v'w'} \\ -\rho\overline{w'u'} & -\rho\overline{w'v'} & -\rho\overline{w'^2} \end{pmatrix}$$

where each τ_{ij} in fact measures the covariance (statistical measure of agreement) between two of the fluctuating components $\mathbf{u}' = (u', v', w')$. If $i = j$ it is the autocovariance of u', v' or w' that is being measured. Both matrices are of course symmetric ($\tau_{ij} = \tau_{ji}$). This follows straight away from the right hand matrix. In fact if $i = j$ then the stress is called a normal stress. Since normal stresses act in a similar fashion to pressure, they can be safely overlooked (or more accurately absorbed by the pressure). When $i \neq j$ the stress is a shear stress and it is these that need to be modelled. It is of course the *gradients* of the stresses that contribute to the conservation of momentum in much the same way as does the gradient of pressure and not the pressure itself. Turbulence and friction have in common the transfer of momentum in a direction at right angles to the flow. The magnitude of the rate at which this transfer takes place in a turbulent ocean current is what is attempted to be modelled by eddy viscosity. The big assumptions are one, that such a relationship is reasonable and two, that the relationship is linear. Neither of these can be justified except perhaps *a posteriori*. To home in on a particular component, let us choose $\tau_{13}(=\tau_{31})$. In this case, the eddy viscosity assumption, a turbulence equivalent to assuming a Newtonian viscous fluid, leads to

$$\tau_{31} = -\rho\overline{w'u'} = -\rho\nu_v \frac{\partial \bar{u}}{\partial z},$$

where ν_v is the eddy viscosity or Austauch coefficient. It is the turbulent equivalent to kinematic (not dynamic) viscosity. Not only is it much bigger than normal viscosity, usually by a factor of 10^6, but it is of course not a fixed property of the fluid. The gradient of the stress and not the stress itself gives the net force. No apology is necessary for reiterating that this is the same as gradients of pressure, not pressure, itself causing net force. Hence if the overall force balance is to include shear stress effects then terms such as

$$\frac{\partial}{\partial z}\tau_{31} = \frac{\partial}{\partial z}\left(\nu_v \frac{\partial u}{\partial z}\right)$$

need to be included. If the eddy viscosity ν_v is a constant, then this can be written in terms of a second derivative:

$$\nu_v \frac{\partial^2 u}{\partial z^2}.$$

The stress τ_{31} is a stress that represents the shear due to a current travelling in an easterly direction in the presence of either the sea bed or the sea surface. Other shear stresses, for example τ_{21} would represent the shear due to an easterly current near a north–south coast, could also be put in terms of gradients in mean flow:

$$\tau_{21} = -\rho\overline{v'u'} = -\rho\nu_H \frac{\partial \bar{u}}{\partial y},$$

but this time the eddy viscosity ν_H is representative of the *horizontal* transfer of momentum in the current. The horizontal eddies that effect this transfer

5. Case Studies

are correspondingly larger than the vertical counterparts, hence $\nu_H \gg \nu_v$ (by a factor of about 10^4 in fact). Let us examine the equation of motion (conservation of momentum) in the x-direction. If we include advection, vertical and horizontal momentum transfer (eddy viscosity) then dimensional analysis can be performed on this equation. As all reference to primed (fluctuating) quantities has ceased, the overbar can be dropped. The general x-wise equation of motion then is rather daunting, until one remembers that is not actually going to be solved. Here it is:

$$\frac{\partial u}{\partial t} + u\frac{\partial u}{\partial x} + v\frac{\partial u}{\partial y} + w\frac{\partial u}{\partial z} - fv = -\frac{1}{\rho}\frac{\partial p}{\partial x} + \nu_v \frac{\partial^2 u}{\partial z^2} + \nu_H \left(\frac{\partial^2 u}{\partial x^2} + \frac{\partial^2 u}{\partial y^2}\right).$$

The y-wise equation is, similarly

$$\frac{\partial v}{\partial t} + u\frac{\partial v}{\partial x} + v\frac{\partial v}{\partial y} + w\frac{\partial v}{\partial z} + fu = -\frac{1}{\rho}\frac{\partial p}{\partial y} + \nu_v \frac{\partial^2 v}{\partial z^2} + \nu_H \left(\frac{\partial^2 v}{\partial x^2} + \frac{\partial^2 v}{\partial y^2}\right).$$

All this is by way of introduction! How much detail is given is of course in the hands of the module or unit leader. Neglecting the terms containing ν_H is equivalent to ignoring horizontal turbulence. If the non-linear terms are ignored too we get

$$\frac{\partial u}{\partial t} - fv = -\frac{1}{\rho}\frac{\partial p}{\partial x} + \nu_v \frac{\partial^2 u}{\partial z^2}$$

$$\frac{\partial v}{\partial t} + fu = -\frac{1}{\rho}\frac{\partial p}{\partial y} + \nu_v \frac{\partial^2 v}{\partial z^2}.$$

These are the unsteady Ekman equations. If we write

$$(u, v) = (u_G + u_E, v_G + v_E)$$

where

$$-fv_G = -\frac{1}{\rho}\frac{\partial p}{\partial x}$$

$$\text{and } fu_G = -\frac{1}{\rho}\frac{\partial p}{\partial y}$$

the equations reduce to

$$\frac{\partial u_E}{\partial t} - fv_E = \nu_v \frac{\partial^2 u_E}{\partial z^2}$$

$$\text{and } \frac{\partial v_E}{\partial t} + fu_E = \nu_v \frac{\partial^2 v_E}{\partial z^2}.$$

We have separated out the geostrophic balance part (suffix G) from the frictional or Ekman part (suffix E). Geostrophic balance is a balance between pressure gradient and Coriolis acceleration which permeates much of the atmosphere and ocean. It gives rise to the characteristic "Lows" and "Highs" of pressure seen on weather maps in temperate regions. The important assumption which allows the above partitioning is that (u_G, v_G) remains independent of z and t. The final assumption is that the Ekman flow is steady, which eliminates the time derivative terms and leaves us with

$$-fv_E = \nu_v \frac{\partial^2 u_E}{\partial z^2}$$
$$\text{and } fu_E = \nu_v \frac{\partial^2 v_E}{\partial z^2}.$$

These can be solved elegantly using complex numbers as follows. We add the first equation to i times the second, where $i = \sqrt{-1}$ of course leading to the single equation

$$ifW = \nu_v \frac{\partial^2 W}{\partial z^2}$$

in the complex quantity $W = u_E + iv_E$. This is a second order differential equation with constant coefficients which can be solved given two boundary conditions on W. One is straightforward. As the wind is the driving force, it is reasonable to suppose that as we get deeper and deeper into the sea, W must get smaller and smaller in magnitude. Thus mathematically

$$W \longrightarrow 0 \text{ as } z \longrightarrow -\infty,$$

remembering that z points upwards. The second boundary condition is more subtle. At the sea surface ($z = 0$) the wind blows. If we assume that the stress imparted by the wind is precisely the same as the shear stress experienced by the sea at the surface, then our previous assumptions concerning eddy viscosity leads to

$$\tau = (\tau^x, \tau^y) = -\rho\nu_v \left(\frac{\partial u_E}{\partial z}, \frac{\partial v_E}{\partial z} \right) = -\rho\nu_v \frac{\partial W}{\partial z},$$

at $z = 0$, where $\tau = (\tau^x, \tau^y) = \tau^x + i\tau^y$ is the (horizontal) wind stress vector. Solving the Ekman equations is now straightforwardly done and gives the solution

$$W = -\frac{\tau}{\rho\sqrt{if\nu_v}} \exp\left\{ z\sqrt{\frac{if}{\nu_v}} \right\}.$$

Having done the hard work and worked through all the assumptions carefully, we can reap the rewards and explore with confidence the properties of this solution. Many of the properties are actually verified by oceanographic observation. At the surface of the sea, $z = 0$ so

$$W = -\frac{\tau}{\rho\sqrt{if\nu_v}} = \frac{\tau}{\rho}\sqrt{\frac{i}{f\nu_v}}$$

and using the properties of complex numbers, it is easy to see that this implies that the surface current is at a 45° angle to the right of the wind in the northern hemisphere. (South of the equator, f is negative which implies the current flows 45° to the left of the wind.) This verifies observations made by Nansen, the Norwegian explorer who stuck his boat the Fram in the ice north of Svålbard (Spitzbergen) in 1896 in a bid to be the first to reach the North Pole. He was there three years and was not successful. In fact it was the request from Nansen to Ekman to explain this angle that led Ekman to his theory that has been repeated here. Ekman published his paper in 1905, the same year as Einstein's seminal publication outlining the special theory of relativity for the first time.

Here are a few auxiliary questions and suggestions for extensions.

Integrate the current through z and show that the net transport is at right angles to the wind direction. Show also that this net transport is independent of the eddy viscosity. (A good student should be able to deduce that the eddy viscosity assumption is unnecessary for this particular result.)

A slightly longer exercise would be to solve the time dependent Ekman equations:

$$\frac{\partial u_E}{\partial t} - f v_E = \nu_v \frac{\partial^2 u_E}{\partial z^2}$$

$$\text{and } \frac{\partial v_E}{\partial t} + f u_E = \nu_v \frac{\partial^2 v_E}{\partial z^2}.$$

This is most easily accomplished using the complex method as earlier, then using Laplace transforms. This leads to an ordinary differential equation which can be solved in terms of z and s (the Laplace transform variable). The solution is given in Laplace transform space as

$$\overline{W} = -\frac{\tau}{\rho\sqrt{(s+if)\nu_v}} \exp\left\{ z\sqrt{\frac{s+if}{\nu_v}} \right\}$$

where the overbar denotes the Laplace transform. It is impossible to invert this solution in closed form in terms of elementary functions. However there are theorems such as the final value theorem and the initial value theorem (see P. P. G. Dyke's excellent Springer-Verlag Undergraduate text: "An Introduction to Laplace Transforms and Fourier Series", published in 2001) which can be used to deduce the spin-up time for the Ekman layer, as the steady solution is called.

Another avenue to explore is the solutions for the steady Ekman equations when the eddy viscosity depends on z. A linear dependence leads to Bessel's differential equation; a quadratic dependence if designed carefully enough can lead to Euler's differential equation with a solution in terms of powers of z. It is interesting to see how this modification changes the angle the surface current makes with the wind. If the students have been given the above material for self study, there are other more adventurous avenues to explore such as the effects of the earth's curvature (the β effect), unsteady effects such as Kelvin or Rossby waves or even a simple model of El Niño. Enough! The author could spend many more pages exploring extensions, but instead let us explore a different case study.

5.3 Non-linear Oscillations

Non-linear oscillations is a subject that has received a considerable boost recently through public interest in chaotic systems and several executive toys have been based on this. It is principally the enhanced power of the computer which has enabled recent advances in the study of dynamic systems, of which non-linear oscillations form a subset. However, the material for this case study originated from work done in the 1970s on designing stable semi-submersibles, ships and moored vessels and so general dynamical systems are not introduced. However, some students may be familiar with phase planes and the properties of solutions to non-linear first order differential equations which form part of the study of dynamic systems. These topics can slot conveniently into a more general course on differential equations. Indeed it is quite common for students to meet the idea of direction fields for non-linear first order differential equations as early as in the first year of a mathematics degree. These topics are also related to the study of characteristics which is introduced briefly in the next section. Characteristics can also be part of an optional course on many undergraduate degrees.

The mathematics of this case study is firmly set against classical solution techniques to ordinary differential equations. Therefore knowledge of power series solutions, singularities, stability and the use of the Wronskian all feature and should be familiar to the students. The vehicle for the study is Duffing's equation with a sinusoidal right hand side which represents non-linear oscillations subject to periodic forcing. Here it is:

$$\ddot{x} + k\dot{x} + \alpha x + \beta x^3 = \Gamma \cos(\omega t).$$

In the last case study, the governing equations were painstakingly derived from physical laws. The derivation of this from Newton's second law of motion is easier and is not done here. Whether or not it is done in class is left to the lecturer's judgement. A Fourier series solution to this is sought, the explanation being that oscillations are expected, but their frequency is not obvious. So set

$$x(t) = a_0 + a_1 \cos(\lambda t) + a_2 \cos(2\lambda t) + a_3 \cos(3\lambda t) + \ldots$$

and substitute into the forced Duffing equation. The x^3 term is still periodic but generates all kinds of cross-terms. If like $\cos(\lambda nt)$ terms are equated on both sides, theoretically the Fourier series solution can be found. The full story can be found in specialist texts such as D.W. Jordan and P. Smith *Non-linear Ordinary Differential Equations* the third edition of which was published in 1999 by Oxford University Press. As a trial solution we could try

$$x(t) = a\cos(\omega t) + b\sin(\omega t).$$

This represents the first two terms of a Fourier series. The equation does not allow a constant term, but this severely truncated Fourier series does allow for phase differences. Substitution and some manipulation of the cube term leads to

$$b\{(\omega^2 - 1) - \frac{3}{4}\beta(a^2 + b^2)\} = 0$$
$$a\{(\omega^2 - 1) - \frac{3}{4}\beta(a^2 + b^2)\} = -\Gamma$$

which implies $b = 0$ so the phase difference disappears! Moreover a has to be a root of the cubic

$$\frac{3}{4}\beta a^3 - (\omega^2 - 1)a - \Gamma = 0.$$

The terms involving trigonometric functions of $3\omega t$ are neglected as they are assumed small compared to the terms having the fundamental frequency ω. These solutions are investigated graphically. For example there are three solutions provided $\beta < 0$, $\omega^2 < 1$ and Γ is small. If $\omega^2 > 1$ there is only ever one solution. The next slightly more general solution to try is not to include further terms of the Fourier series, but to allow the coefficients a and b to become $a(t)$ and $b(t)$, i.e. slowly varying functions of time. Once more ignoring the $3\omega t$ harmonics this time leads to the coupled first order system:

$$\dot{a} = -\frac{b}{2\omega}\{(\omega^2 - 1) - \frac{3}{4}\beta(a^2 + b^2)\}$$
$$\dot{b} = \frac{a}{2\omega}\{(\omega^2 - 1) - \frac{3}{4}\beta(a^2 + b^2)\} + \frac{\Gamma}{2\omega}$$

which in the language of systems represents an *autonomous* system. Solutions are drawn in the (a, b) plane, and equilibrium points $\dot{a} = 0$ and $\dot{b} = 0$ represent the constant a and b solutions already obtained. These equilibrium points can be analysed in terms of their stability. As mentioned earlier, there can be three such points or there could be just one depending on the particular values of the parameters. This is a rich source of mathematical problems and the case study could, at this juncture, explore many possibilities. Chapter 7 of Jordan and Smith (1999) does just this.

Another route is to explore oscillations not at the frequency of the forcing but at some fraction of this frequency. These are called subharmonic oscillations and they can have important consequences for design engineers. Of course a simple linear SHM equation such as

$$\ddot{x} + \frac{1}{n^2}x = \Gamma \cos t$$

will have the subharmonic oscillatory solution

$$a \cos \frac{1}{n}t + b \sin \frac{1}{n}t - \frac{\Gamma}{1 - n^{-2}} \cos t,$$

but subharmonics also occur less obviously. Looking for perturbation solutions of Duffing's Equation

$$\ddot{x} + k\dot{x} + \alpha x + \beta x^3 = \Gamma \cos(\omega t)$$

regarding β as a small parameter is constructive. Proceed by letting $\tau = \omega t$ which gives

$$\omega^2 x'' + k\omega x' + \alpha x + \beta x^3 = \Gamma \cos(\tau),$$

where primes denote differentiation with respect to τ. The perturbation method is now applied in which the solution $x(\tau)$ is expanded in a power series in β

$$x(\tau) = x_0(\tau) + \beta x_1(\tau) + \beta^2 x_2(\tau) + \ldots$$

as is ω

$$\omega = \omega_0 + \beta \omega_1 + \beta^2 \omega_2 + \ldots.$$

Taking damping $k = 0$ for simplicity, this leads to a series of linear differential equations, the first two of which are

$$\omega_0^2 x_0'' + \alpha x_0 = \Gamma \cos(\tau)$$

and

$$\omega_0^2 x_1'' + \alpha x_1 = -2\omega_0 \omega_1 x_0'' - x_0^3.$$

In order for there to be periodic solutions to this last equation we must have

$$\frac{\alpha}{\omega_0^2} = \frac{1}{9} \quad \text{or} \quad \omega_0 = 3\sqrt{\alpha}.$$

This immediately gives the following: a subharmonic of one third is stimulated by applied forcing at three times the natural frequency around $\beta = 0$. The presence of these subharmonics is therefore to be expected if the system is well approximated by Duffing's equation and the non-linearity can be regarded as a perturbation from the linear. The solution for $x_0(\tau)$ is

$$x_0(\tau) = a_{1/3} \cos \frac{1}{3}\tau + b_{1/3} \sin \frac{1}{3}\tau - \frac{\Gamma}{8\alpha} \cos(\tau).$$

The constants $a_{1/3}$ and $b_{1/3}$ are evaluated by demanding that the right hand side of the equation for $x_1(\tau)$ has to be zero so that x_1 is periodic and free of resonance. This gives non-linear equations for these constants, and the perturbation solution is built up in this way. It is important in case studies such as this to emphasise the practical application. In this case one can discuss oscillations of floating platforms due to waves and how the unfortunate presence of waves of a certain frequency could excite subharmonics present in the platforms and cause the platform to behave violently and even overturn. Stability analysis can be effectively employed on this perturbation solution: however there is yet a third way to analyse Duffing's equation.

First of all Duffing's equation is transformed into a coupled system of first order ordinary differential equations by setting $y = \dot{x}$. Hence the undamped version

$$\ddot{x} + x + \beta x^3 = \Gamma \cos(\omega t)$$

where we have set $\alpha = 1$ becomes

$$\begin{aligned} \dot{x} &= y \\ \dot{y} &= -x - \beta x^3 + \Gamma \cos \omega t. \end{aligned}$$

The method then proceeds using variational arguments. Suppose

$$\begin{aligned} \xi &= x - x^* \\ \eta &= y - y^* \end{aligned}$$

and we substitute for x and y into the above equations but ignore squares and higher powers of ξ and η. Thus

$$\begin{aligned} \dot{\xi} &= \eta \\ \dot{\eta} &= -\xi - 3\beta x^{*2}\xi \end{aligned}$$

as $\dot{x}^* = \dot{y}$ and $\dot{y}^* = -x^* - \beta x^{*3} + \Gamma \cos \omega t$. Periodic solutions of the form

$$x = a \cos \omega t$$

are now sought. In order for this to *be* a solution, we need

$$\frac{3}{4}\beta a^3 + (1 - \omega^2)a = \Gamma.$$

This is obtained by substituting into the original Duffing's equation for x and equating coefficients of $\cos \omega t$. Again this is justified as long as $a \cos \omega t$ is an approximate solution which is in effect the first term of a Fourier series. If we eliminate η from the above pair of equations in ξ and η, then

$$\ddot{\xi} + (1 + 3\beta x^{*2})\xi = 0.$$

Further we approximate x^* by $a\cos\omega t$ with a given by the root of the cubic above, to give the following equation for ξ:

$$\ddot{\xi} + (1 + \frac{3}{2}\beta a^2 + \frac{3}{2}\beta a^2 \cos 2\omega t)\xi = 0.$$

This enables the stability of the solution to be analysed. This equation is in fact a form of Mathieu's equation which arises naturally from the equation obeyed by a simple pendulum whose support oscillates vertically. One possible direction for an extension to this case study is to derive the equation to such a pendulum in the form

$$a\ddot{\theta} + (g - \ddot{z})\theta = 0$$

where a is the length of the pendulum, z the amplitude of the vertical oscillation of its support and θ the small angle the pendulum string makes with the vertical. As the coefficient of θ in the above equation is periodic it is Hill's equation

$$\ddot{x} + (\alpha + p(t))x = 0$$

which becomes Mathieu's equation for the case when $p(t) = k\cos\omega t$. This is an example of *parametric excitation* which is behind many engineering problems including the oscillation of the Millennium Bridge over the Thames in London.

There are many other possible extensions of this case study. One could be the analysis of subharmonic oscillations of floating bodies, including the generalisation to n dimensional linear systems (Floquet theory) and the inclusion of damping. Another is the analysis of N linked forced pendulums. This has recently been examined by D. J. Acheson (see his papers in the **Proceedings of the Royal Society of London**, 1993, **443** 239–245, and 1995, **448** 89–95.) This derives Mathieu's equation, and relates the stability to parameter ranges. This particular approach is elegant but quite straightforward and lends itself to investigation by a final year student. Of course, if the degree explicitly contains modules on non-linear dynamics and chaos, many more extensions are possible.

5.4 Traffic Flow

It may not be immediately obvious, but this case study stems from the study of wave motion. Many mathematics degrees have something in them about waves, but very few have dedicated courses that are compulsory. This is a common problem in introducing case studies: one has to start from a common knowledge base. To assume all students know about linear wave theory would discourage too many students, but to give a full account of linear wave theory would not be possible. A compromise is to give a quick run through of linear waves, in

particular to introduce the terminology. The simplest description centres on the sinusoidal wave
$$\phi(x,t) = a\sin(kx - \omega t).$$
This is a wave of *amplitude a*, *wave number k*, *angular frequency* ω, propagating along the x-axis. t represents time. The *frequency* is $f = \omega/2\pi$ (in Hertz or cycles per second, whereas angular frequency is measured in radians per second). The *wavelength* is $\lambda = 2\pi/k$, and the *period* is $1/f$ or $2\pi/\omega$. Waves have *crests* and *troughs* corresponding to the maxima and minima respectively of the sine wave. *Progressive* waves retain their form (sinusoidal above, but any periodic shape repeated indefinitely will do). Such a wave moves with *celerity* (alternative names are *phase velocity* or *wave speed*) $c = \omega/k$ along the x-axis. The other extreme is a wave that does not progress at all and is a *standing* wave. An example with a sinusoidal profile would be
$$\phi(x,t) = a\sin kx \cos \omega t.$$
A standing wave has points (*nodes*) where there is never any vertical displacement. For the above wave these occur at $x = \frac{\pi}{k}n$ where $n = 0, \pm 1, \pm 2, \ldots$. At times $t = \frac{\pi}{2\omega}(2N+1)$ where $N = 0, \pm 1, \pm 2, \ldots$, the wave is completely flat. Most waves in reality are someway between being entirely progressive or entirely standing. All waves can be expressed as a train of progressive waves. The standing wave above is simply two opposing progressive waves:
$$\frac{1}{2}(a\sin(kx - \omega t) + a\sin(kx + \omega t)).$$

The differential equation obeyed by the one-dimensional wave (progressive or standing or anything in between) is
$$\frac{\partial^2 \phi}{\partial x^2} = \frac{1}{c^2}\frac{\partial^2 \phi}{\partial t^2}.$$
The variety of solutions to this, especially in three dimensional form
$$\nabla^2 \phi = \frac{1}{c^2}\frac{\partial^2 \phi}{\partial t^2}$$
is vast, and the subject of thick textbooks and year long lecture courses.

Linear waves are met in electromagnetism as solutions to Maxwell's equations: they are also an important solution to the equations of elasticity. In this guise they help us to understand earthquakes. In water they are solutions to the Navier–Stokes' equations met in rotating form earlier in this chapter. Linear water wave theory works well to describe tides, swell waves and other long waves where the wavelength is at least double the wave height (wave height is an engineering term equal to twice the amplitude of a sine wave). However an underwater earthquake can cause a tsunami (once poorly termed "tidal wave" in some old newspaper headlines) which can be virtually a wall of water when it reaches a coast. A tsunami in this situation is not described by linear wave theory. Non-linear waves exhibit features that are very distinctive, amongst which are *shocks*. Perhaps amongst the most familiar shock wave is the bang that accompanies aircraft that break the sound barrier, while the tidal bore is another. Believe it or not, the mathematics of shock waves is the same that can be used to describe the propagation of stoppages in dense traffic. Anyone who regularly uses the M25 ring road around London in the UK will have practical experience of this type of shock wave. This case study describes the mathematics required to understand it.

The crucial property that all linear waves possess is that if ϕ_1 and ϕ_2 are solutions to the wave equation, then so is $a_1\phi_1 + a_2\phi_2$ where a_1 and a_2 are constants. This is no longer the case for non-linear waves and new methods have to be developed to study them. One of the most profitable is the *Method of Characteristics* and this will have to be outlined in order to understand traffic flow modelling. A characteristic curve carries information forward in time from an initial condition. The general theory belongs squarely in courses on partial differential equations, but a few hints for students not familiar with this can run along the following lines. Given a first order differential equation in the form

$$a\frac{\partial u}{\partial x} + b\frac{\partial u}{\partial t} = c,$$

we can write

$$\frac{dx}{dt} = a \qquad \frac{dy}{dt} = b$$

then

$$\frac{dx}{dt}\frac{\partial u}{\partial x} + \frac{dy}{dt}\frac{\partial u}{\partial y} = \frac{du}{dt} = c$$

and the family of characteristic curves is given by $u = u(x, y)$ for which the integrals of these three equations are the parametric form. Although this is linear and first order, this method of finding a family of curves or indeed surfaces that satisfy a partial differential equation generalises to some higher order non-linear equations. It gives a good way of determining solutions, and the phrase "Method of Characteristics" is used for all such methods. Here is an example of determining a solution to a non-linear second order equation using the method of characteristics that could form part of an introduction for students. Consider the equation:

$$t\frac{\partial^2 u}{\partial x^2} + (t+x)\frac{\partial^2 u}{\partial x \partial t} + x\frac{\partial^2 u}{\partial t^2} = 0.$$

The initial values of u and its derivatives are known, and the equation thus enables all derivatives of u to be determined at time $t = 0$. Taylor's theorem can thus be used to find a power series solution. Of course, we cannot solve this equation without knowing these initial conditions: however as

$$\frac{\partial u}{\partial x}(0, x) = f(x) \quad \text{(say)} \quad \text{and} \quad \frac{\partial u}{\partial t}(0, x) = g(x) \quad \text{(say)}$$

we can differentiate these and obtain three equations obeyed by the three second order partial derivatives of u, namely

$$t\frac{\partial^2 u}{\partial x^2} + (t+x)\frac{\partial^2 u}{\partial x \partial t} + x\frac{\partial^2 u}{\partial t^2} = 0$$

$$\frac{\partial^2 u}{\partial x^2}\frac{dx}{dt} + \frac{\partial^2 u}{\partial t \partial x} = \frac{df}{dt} = \frac{df}{dx}\frac{dx}{dt}$$

$$\frac{\partial^2 u}{\partial t \partial x}\frac{dx}{dt} + \frac{\partial^2 u}{\partial t^2} = \frac{dg}{dt} = \frac{dg}{dx}\frac{dx}{dt}.$$

The consistency condition for these three equations is that the determinant of the left hand side is zero. This means that

$$\begin{vmatrix} t & t+x & x \\ \frac{dx}{dt} & 1 & 0 \\ 0 & \frac{dx}{dt} & 1 \end{vmatrix} = 0$$

or

$$x\left(\frac{dx}{dt}\right)^2 - (x+t)\frac{dx}{dt} + t = 0.$$

The solution of this quadratic equation is

$$\frac{dx}{dt} = \frac{t}{x} \quad \text{and} \quad \frac{dx}{dt} = 1.$$

Integrating these gives the two sets of characteristic curves

$$x^2 - t^2 = \text{const.} \quad \text{and} \quad x - t = \text{const.}$$

There is no need to take this particular example further, but knowledge of these characteristic curves enables the equation to be transformed into so-called canonical form which in turn enables the solution to be determined. For fully non-linear equations, the method of characteristics can still be applied: however the characteristics themselves cannot in general be found explicitly. The equation will for example usually be implicit and depend on u. Also, characteristics are only real and distinct for some equations (designated as *hyperbolic partial differential equations*). For equations with variable coefficients, there may be regions where the equation flips to being parabolic and then elliptic in which case the characteristics become complex and cease to exist in real space.

All this can be mentioned as an introduction to this case study, and reference should be made to texts on partial differential equations for further study and perhaps some student exercises.

Turning to traffic flow modelling, we define a single one-dimensional stream of cars. Assume that a car density ρ can be defined and that the local car velocity $v(x,t)$ is a function only of this density, i.e. $v = v(\rho)$. This is a candidate for modification to develop more sophisticated models as it implies that the speed of any given car is entirely determined by the speed of the car in front. If $q(x,t)$ is the car flow rate, then over a length of road $x_1 - x_2$ the conservation of cars implies that

$$\frac{\partial}{\partial t}\int_{x_2}^{x_1} \rho(x,t)dx = q(x_1) - q(x_2).$$

Taking the limit as $(x_1 - x_2) \to 0$ we get

$$\frac{\partial \rho}{\partial t} + \frac{\partial q}{\partial x} = 0$$

which is the one-dimensional version of the continuity equation familiar to those who study fluid mechanics. A certain amount of discussion can be based around the consequences of this equation. For example, given that $q = \rho v$, stationary traffic $v = 0$ imposes an upper limit on the density of cars. Also as density increases, drivers must slow down. Both of these are reasonable, but the model does have shortcomings and does not predict the observation that the maximum flow rate in heavy traffic tends to occur when cars are travelling at moderate speeds. Go too quickly and a traffic jam results. A traffic jam where the cars are stationary can be analysed by using wave theory coupled with characteristics. The outline here owes much to the excellent account given in *Wave Motion* by J. Billingham and A. C. King, published by Cambridge University Press in 2000. The problem is expressed mathematically as follows:

$$\frac{\partial \rho}{\partial t} + c(\rho)\frac{\partial \rho}{\partial x} = 0$$

where

$$c(\rho) = \frac{d}{d\rho}(\rho v) = v(\rho) + \rho v'(\rho)$$

is the kinematic wave speed. This name stems from the solution for small variations in density for which the problem is linear. In this case, the system is perturbed from a state $\rho = \rho_0$ so that $\rho = \rho_0 + \tilde{\rho}$ where $\tilde{\rho} << 1$. The above equation then approximates to the one-dimensional linear equation

$$\frac{\partial \tilde{\rho}}{\partial t} + c(\rho_0)\frac{\partial \tilde{\rho}}{\partial x} = 0$$

which has a general solution:
$$\tilde{\rho} = f(x - c(\rho_0)t)$$
representing a wave propagating with speed (celerity) $c(\rho_0)$. This is rather idealised, but it does tell us that should there be a maximum value of the flux $q(\rho)$ at $\rho = \rho*$ say, then $c > 0$ for $\rho < \rho*$ but $c < 0$ for $\rho > \rho*$. This is the phenomenon of the traffic wave travelling in the opposite direction to the flow and corresponds to the traffic ahead stopping for no apparent reason and this stopped traffic backing up along the road. Curves upon which the traffic density is constant are the characteristic curves for this problem in $x-t$ space. For this linear approximation, the characteristic curves are simply the straight lines
$$x = x_0 + c(\rho_0)t$$
and these are shown in Figure 5.2. The non-linear problem brings out the value

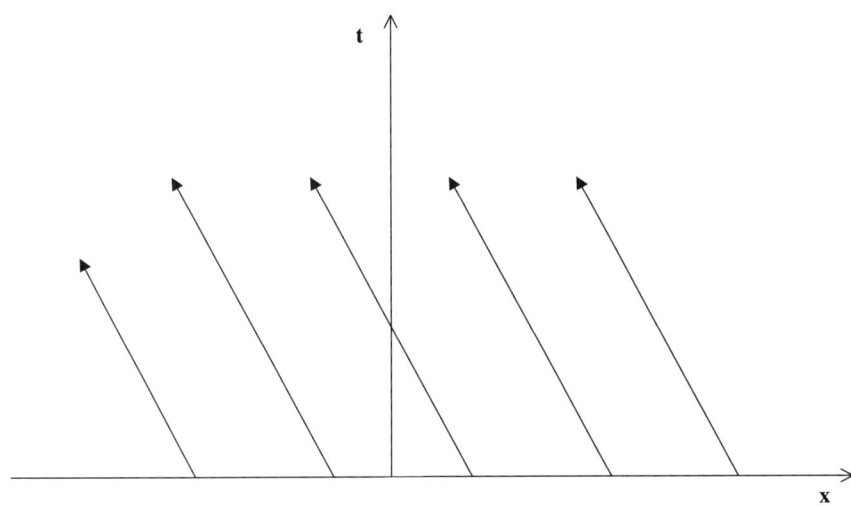

Figure 5.2   The characteristics for the linear approximation.

of using characteristics. In order to focus a case study, it is a good idea to examine a specific problem. Let us set as car velocity
$$v(\rho) = \frac{v_0}{\rho_{max}}(\rho_{max} - \rho)$$

from which the celerity is

$$c(\rho) = \frac{d}{d\rho}(\rho v) = \frac{v_0}{\rho_{max}}(\rho_{max} - 2\rho),$$

and a specific functional form for the initial car density is chosen as follows:

$$\rho_0(x) = \frac{\rho_L + \rho_R e^{x/L}}{1 + e^{x/L}},$$

where ρ_L, ρ_R and L are positive constants. This is a representation of density changing gradually from ρ_L for large but negative x to ρ_R for large and positive x crossing the value $\rho_0 = L$ at $x = 0$. The case where $\rho_L > \rho_R$, which will always be assumed here, is shown in Figure 5.3. Now, we are considering the non-linear

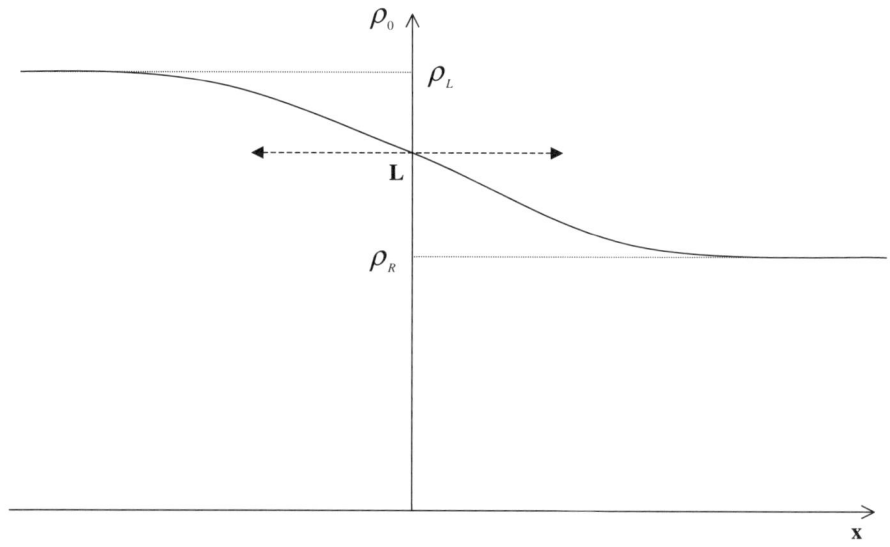

Figure 5.3 The initial conditions for car density $\rho(x, 0) = \rho_0(x)$.

problem:

$$\frac{\partial \rho}{\partial t} + c(\rho)\frac{\partial \rho}{\partial x} = 0$$

subject to

$$\rho(x, 0) = \rho_0(x)$$

which has been assigned above. The task is to find a set of characteristics upon which to base the solution to the boundary value problem. As the problem is driven by the initial conditions, from the properties of characteristics outlined above, we seek a solution $\rho(x,t)$ as follows. Look for a characteristic curve $x = X(t)$ for which

$$\rho(X(t),t) = \rho(X(0),0) = \rho_0(X(0)).$$

That is, differentiating as in the above example on characteristics,

$$\frac{d}{dt}(\rho(X(t),t)) = \frac{\partial \rho}{\partial t} + \frac{dX}{dt}\frac{\partial \rho}{\partial x}.$$

Having defined the celerity $c(\rho)$ through

$$\frac{d}{dt}(\rho(X(t),t)) = \frac{\partial \rho}{\partial t} + c(\rho)\frac{\partial \rho}{\partial x}$$

it is clear that we require

$$c(\rho) = \frac{dX}{dt}.$$

By construction, ρ is a constant on each characteristic, therefore so is $c(\rho)$. Hence the above equation implies that, even in this non-linear case, each characteristic remains a straight line and takes the form

$$x = X(t) = x_0 + c(\rho_0(x_0))t \qquad \text{for} \qquad -\infty < x_0 < \infty.$$

Inserting the chosen form for $\rho_0(x)$ into these expressions gives explicit formulae for the characteristics. Several student exercises lend themselves to the lecturer at this point. First of all, students should be able to draw the graph of $c(\rho)$ as a function of x. Next, the characteristic curves (straight lines) should be drawn on an $x-t$ plot. They will take the form of a "fan". Students could be asked to explore numerical solutions to the boundary value problem too. The following solution to the problem is well worth exploring as it gives several insights:

$$\rho(x,t) = \begin{cases} \rho_L & \text{for } x < c(\rho_L)t \\ \frac{1}{2}\rho_{max}(1 - \frac{x}{v_0 t}) & \text{for } c(\rho_L)t \leq x \leq c(\rho_R)t \\ \rho_R & \text{for } x \geq c(\rho_R)t. \end{cases}$$

For example, with $\rho_L = \rho_{max}$ and $\rho_R = 0$ cars pulling away from a queue at traffic lights as they turn green is simulated. The "shock", that is the interface between stationary and moving cars, propagates back through the queue with celerity $c(\rho_{max}) = -v_0$. The general solution curves for $\rho(x,t)$ should be drawn against $x-c(\rho_0(0))$, and the gradual dissipation of the front at $x = c(\rho_0(0))$ with time noted. Other generalisations of a mathematical nature are possible (see the book by Billingham and King referred to earlier) as well as observations such as if a traffic jam is approached at a slow enough speed it will always vanish, which is obvious after a little thought. Conversely, fast driving in dense traffic causes jams, also obvious although you wouldn't think so given the evidence!

Students can be asked to take this case study further. If the taught part of their course contains the theory of characteristics, then they should be able to analyse the characteristics defined here more fully. For example, investigate what is represented by their intersection and interpret this in terms of shock waves. In turn, this can be interpreted in terms of the behaviour of real traffic, and when it cannot students can investigate which of the initial assumptions are violated. Students can also investigate alternative functional forms for the traffic density ρ.

5.5 Contour Integral Solutions to ODEs

Back in the mists of time, a lecture course on complex variables formed a compulsory part of the first two years of any self respecting mathematics degree. This is no longer the case, and it is not only possible but commonplace for mathematics students to graduate knowing no complex variable theory at all, and little in the way of related matter beyond the arithmetic of complex numbers. Such students are denied the use of the power of complex variables. A particularly strong use is in finding solutions to both ordinary and partial differential equations and it is this that forms this case study.

As a prerequisite, students should know how to solve both ordinary and partial differential equations. Solution methods such as the Frobenius method, variation of parameters and complementary function and integrating factor techniques for ordinary differential equations, and the use of separation of variables should be included. Some knowledge of Laplace transforms would also be useful, but not essential.

As this case study concentrates on finding contour integral solutions to certain types of ordinary differential equation, so a knowledge of complex variable theory up to the definition of integration and singularities is essential. In fact if the students have not had the opportunity to study a unit on complex variable theory, this case study is unsuitable.

To begin in the most general way, consider the ordinary differential equation of nth order but with linear coefficients. This is sometimes referred to as Laplace's linear equation, and it is written in the form

$$(a_n z + b_n)\frac{d^n w}{dz^n} + (a_{n-1} z + b_{n-1})\frac{d^{n-1} w}{dz^{n-1}} + \ldots + (a_0 z + b_0)w = 0,$$

where a_n and b_n are (complex) constants, $z = x + iy$ is the standard complex variable and $w = w(z)$ is a function of the complex variable z. The key point here is that with linear coefficients, an integral transform technique works. If the coefficients contain quadratic or higher powers of z there is no guarantee that the method will work, though it might as we will see in the last example in this section. Those students familiar with the technique of using Laplace transforms for solving ordinary differential equations (see P. P. G. Dyke's *Introduction to Laplace Transforms and Fourier Series*, published by Springer-Verlag in 2001, chapter 3) will have a head start. Those for whom the technique is new will need a little more time. The idea is to set

$$w = \int_C f(z, \zeta) d\zeta$$

where $f(z, \zeta)$ is a complex valued function of both z and ζ. The behaviour of f in the ζ plane (for fixed z) is such that it has a finite number of isolated singularities in the form of poles and/or branch points but not essential singularities. This means that along any particular branch of the contour C if the function $f(z, \zeta)$ contains the factor $(\zeta - a)^k$ (real k) then although ostensibly this is many valued a value may be assumed as long as we remain on a chosen branch of the function. There are many concepts here; singularity, pole, branch point, essential singularity. It is therefore reinforced that it is a good idea for this case study to take place after a reasonable grounding in complex variable theory.

The way to proceed is already hinted at with the mention of Laplace Transforms. We try setting

$$f(z, \zeta) = e^{z\zeta} P(\zeta)$$

so that

$$w = \int_C e^{z\zeta} P(\zeta) d\zeta.$$

First of all, since it is legitimate to exchange derivative and integral operations for this integrand, we note that

$$\begin{aligned} b_p \frac{d^p w}{dz^p} &= b_p \frac{d^p}{dz^p} \left\{ \int_C e^{z\zeta} P(\zeta) d\zeta \right\} \\ &= b_p \int_C \zeta^p e^{z\zeta} P(\zeta) d\zeta, \end{aligned}$$

for $p = 1, 2, \ldots, n$. This is the "Laplace transform trick". The terms in a_p need more careful attention as follows:

$$\begin{aligned} a_p z \frac{d^p w}{dz^p} &= a_p z \frac{d^p}{dz^p} \left\{ \int_C e^{z\zeta} P(\zeta) d\zeta \right\} \\ &= a_p \int_C z \zeta^p e^{z\zeta} P(\zeta) d\zeta \\ &= a_p \left[\zeta^p e^{z\zeta} P(\zeta) \right]_C - a_p \int_C e^{z\zeta} \frac{d}{d\zeta} \left(\zeta^p P(\zeta) \right) d\zeta. \end{aligned}$$

Thus in order to satisfy Laplace's linear equation we require that

$$\sum_{p=0}^{n} \left\{ b_p \int_C \zeta^p e^{z\zeta} P(\zeta) d\zeta + a_p \left[\zeta^p e^{z\zeta} P(\zeta)\right]_C - a_p \int_C e^{z\zeta} \frac{d}{d\zeta}\left(\zeta^p P(\zeta)\right) d\zeta \right\} = 0.$$

This can be written

$$\left[A(\zeta)P(\zeta)e^{z\zeta}\right]_C + \int_C \left[B(\zeta)e^{z\zeta}P(\zeta) - e^{z\zeta}\frac{d}{d\zeta}(A(\zeta)P(\zeta))\right] d\zeta = 0$$

where

$$A(\zeta) = \sum_{p=0}^{n} a_p \zeta^p$$

and

$$B(\zeta) = \sum_{p=0}^{n} b_p \zeta^p.$$

This is satisfied by setting

$$B(\zeta)P(\zeta) = \frac{d}{d\zeta}(A(\zeta)P(\zeta))$$

and by demanding that

$$\left[A(\zeta)P(\zeta)e^{z\zeta}\right]_C = 0.$$

The first order differential equation for $P(\zeta)$ that results from the first of these demands leads to $P(\zeta)$ taking the form:

$$P(\zeta) = e^{k_0 \zeta}(\zeta - \alpha_1)^{k_1} \ldots (\zeta - \alpha_n)^{k_n}$$

where $\alpha_1, \ldots, \alpha_n$ are the zeros of the polynomial $A(\zeta)$ provided they are all different. The requirement that the integrated part vanishes takes the form

$$\left[e^{(z+k_0)\zeta}(\zeta - \alpha_1)^{(k_1+1)} \ldots (\zeta - \alpha_n)^{k_n+1)}\right]_C = 0$$

which leads to choices for the contour C which are both interesting and informative. This general theory is a good exercise for the students in the use and interpretation of complex contour integration. However specific examples should be done in class in order to find specific $f(z,\zeta)$s and specific contours. A good choice is to solve Airy's differential equation. This equation occurs in several branches of applied mathematics where an oscillatory solution in one part of the solution domain gives way to an exponential solution in another. Early models of the steady currents that inhabit equatorial ocean regions are one example. Airy's ordinary differential equation takes the form

$$w'' = zw.$$

5. Case Studies

Let
$$w = \int_C e^{z\zeta} P(\zeta) d\zeta$$
so on differentiating twice with respect to z
$$w'' = \int_C \zeta^2 e^{z\zeta} P(\zeta) d\zeta$$
and
$$\begin{aligned} wz &= \int_C z e^{z\zeta} P(\zeta) d\zeta \\ &= \left[e^{z\zeta} P(\zeta)\right]_C - \int_C e^{z\zeta} P'(\zeta) d\zeta. \end{aligned}$$

So Airy's equation $w'' = zw$ implies
$$\left[e^{z\zeta} P(\zeta)\right]_C - \int_C (\zeta^2 P(\zeta) + P'(\zeta)) e^{z\zeta} d\zeta = 0.$$

To satisfy this, we first choose the function $P(\zeta)$ to satisfy an equation that makes the second expression vanish, viz.
$$-\zeta^2 P(\zeta) = P'(\zeta)$$
from which on integration (ignoring the arbitrary constant–we only want *one* solution) gives
$$P(\zeta) = e^{-\frac{1}{3}\zeta^3}$$
together with demanding that
$$\left[e^{z\zeta} P(\zeta)\right]_C = \left[e^{z\zeta - \frac{1}{3}\zeta^3}\right]_C = 0.$$

Now, the $-\frac{1}{3}\zeta^3$ part will dominate far from the origin. Also, writing $\zeta = re^{i\theta}$ we have
$$-\frac{1}{3}\zeta^3 = -\frac{1}{3}r^3 \cos 3\theta - \frac{i}{3}r^3 \sin 3\theta$$
and the real part of this is negative provided
$$\cos 3\theta > 0.$$

This is the case if
$$-\frac{\pi}{2} < 3\theta < \frac{\pi}{2}; \qquad \frac{3\pi}{2} < 3\theta < \frac{5\pi}{2}; \qquad -\frac{5\pi}{2} < 3\theta < -\frac{3\pi}{2},$$
or
$$-\frac{\pi}{6} < \theta < \frac{\pi}{6}; \qquad \frac{\pi}{2} < \theta < \frac{5\pi}{6}; \qquad -\frac{5\pi}{6} < \theta < -\frac{\pi}{2}.$$
One such contour is illustrated in Figure 5.4. Here it is possible to choose two

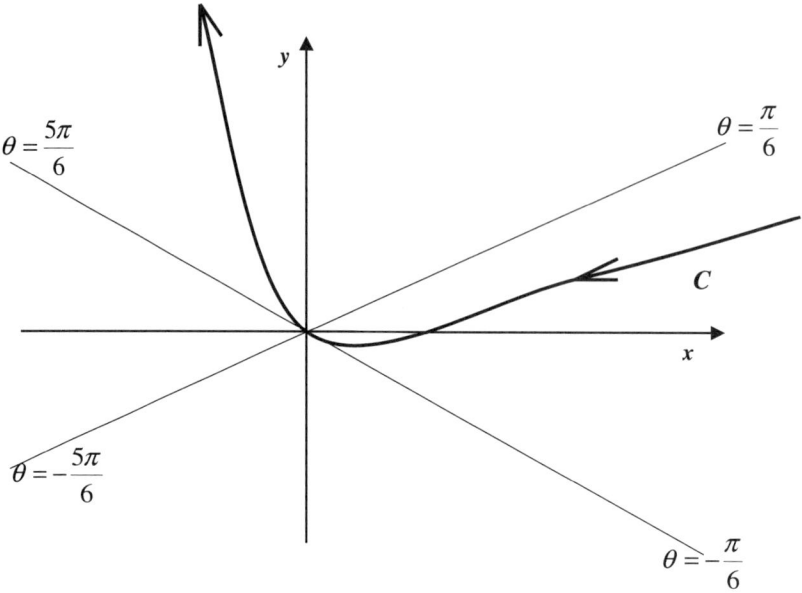

Figure 5.4 One solution contour for Airy's equation.

(but not three) independent contours, that is contours that are not deformable into each other. This illustrates that there are indeed two independent solutions to this second order differential equation. One possible extension to set the students is to solve the equation

$$w''' = zw$$

or perhaps the generalised Airy equation

$$w^{(n)} = zw.$$

Neither of these would tax the good student. There are other equations one could tackle by this method, for example the Legendre equation:

$$(1-z^2)\frac{d^2w}{dz^2} - 2z\frac{dw}{dz} + n(n+1)w = 0.$$

This is not a member of Laplace's linear ordinary differential equations: nevertheless an contour integral solution can be obtained as follows. A power series solution can be generated in the standard form

$$w = z^{-\rho}\sum_{r=0}^{\infty} c_r z^{-r}.$$

This is because $z = \infty$ is one of the three regular singularities of this differential equation (the others are at $z = \pm 1$). Standard series solution methods are used to find the c_r and these may need to be revised. A polynomial solution is possible provided n is an integer. It has the form

$$w = Az^n \left\{ 1 - \frac{n(n-1)}{2(2n-1)} z^{-2} + \frac{n(n-1)(n-2)(n-3)}{2.4.(2n-1)(2n-3)} z^{-4} - \cdots \right\}.$$

Taking the constant A to have the value

$$\frac{(2n)!}{2^n (n!)^2}$$

in turn leads to the Rodrigue's formula for the Legendre equation

$$\frac{1}{2^n n!} \frac{d^n}{dz^n} (z^2 - 1)^n.$$

This is a neat way of writing the polynomial solution which students may have met before. If they have, so much the better. These days, the next bit will probably be new. The Cauchy integral formula

$$\frac{d^n}{dz^n} f(z) = \frac{n!}{2\pi i} \int_C \frac{f(\zeta)}{(\zeta - z)^{n+1}} d\zeta$$

is introduced. Putting

$$f(z) = \frac{(z^2 - 1)^n}{2^n n!}$$

we get the contour integral solution

$$\frac{1}{2\pi i} \int_C \frac{(\zeta^2 - 1)^n}{2^n (\zeta - z)^{n+1}} d\zeta.$$

Such integrals are called Schläfli integrals. Having derived this using n as an integer we can allow n to be completely general and rename it ν to emphasise this. The Schläfli integral remains a solution as can be seen by inserting the integral

$$w = \frac{1}{2\pi i} \int_C \frac{(\zeta^2 - 1)^\nu}{2^\nu (\zeta - z)^{\nu+1}} d\zeta$$

into Legendre's equation

$$(1 - z^2) \frac{d^2 w}{dz^2} - 2z \frac{dw}{dz} + \nu(\nu + 1) w = 0$$

but we need to find C such that

$$\frac{\nu + 1}{2\pi i 2^\nu} \left[\frac{(\zeta^2 - 1)^{\nu+1}}{(\zeta - z)^{\nu+2}} \right]_C = 0.$$

Of course, since now ν is no longer an integer, the three points $\zeta = 1$, $\zeta = -1$ and $\zeta = z$ are all branch points so this condition is no longer automatically satisfied. However, by choosing the contour C with enough care, solutions to Legendre's differential equation are obtained. A simple closed contour containing the points $\zeta = 1$, $\zeta = z$ but not $\zeta = -1$ gives one solution. This works because $\nu + 1 - (\nu + 2)$ is an integer. Another possible choice for C is a figure of eight around $\zeta = 1$ and $\zeta = -1$ not containing the point $\zeta = z$. Again such a contour works since $\zeta = 1$ and $\zeta = -1$ are circled in opposite senses which means that the value of
$$\frac{(\zeta^2 - 1)^{\nu+1}}{(\zeta - z)^{\nu+2}}$$
does not change as the contour is traversed. Additionally the two contours are not deformable into each other; hence we have two independent solutions. There are no others.

The students could be given Laguerre's equation, Bessel's equation or some other equation with a polynomial solution and be required to go through similar steps as above to obtain a Schläfli integral solution. The classic text: *A Course of Modern Analysis* by E. T. Whittaker and G. N. Watson although old (1902) is a rich source for examples.

This case study will only be accessible to students with a reasonable knowledge of complex variable theory. However, it does show the power of complex variables to generate solutions to ordinary differential equations. Other directions to pursue are the relationship of Schläfli integrals to generating functions, and the use of similar methods to solve partial differential equations . This latter topic is probably more suited to MMath, MSci or other advanced mathematics courses.

5.6 Optimisation

For the past thirty five years undergraduate degrees in mathematics have contained some numerical analysis or at least some numerical methods. This of course reflects the developments in computing. These days, there are sophisticated software packages that enable students to pursue many different kinds of numerical mathematics. Software such as MAPLE are designed to minimise the amount of computer programming, an unpopular topic with students (especially computer science students oddly enough). This section gives one area that students can explore using numerical mathematics.

Optimisation is a large and expanding area of mathematics concerned with finding the "best". This might be the maximum value of profit, the minimum value of cost, the best route through, the procedure that gets there fastest, etc. Mathematically all of these require finding a maximum or minimum of a function, and this in turn requires finding the roots (solutions) of an equation. In one dimension this takes the form

$$f(x) = 0.$$

Most students will know more than one method for solving such an equation numerically. Perhaps the most common is the Newton–Raphson method. This and other cruder methods such as bisection are met in the first year. Indeed, solving $f(x) = 0$ numerically is a popular vehicle for coursework in 'A' level (A2) Mathematics taken during the last year at school. In many dimensions this optimisation problem generalises to

$$\begin{aligned} f_1(x_1, x_2, \ldots, x_n) &= 0 \\ f_2(x_1, x_2, \ldots, x_n) &= 0 \\ f_3(x_1, x_2, \ldots, x_n) &= 0 \\ &\vdots \\ f_n(x_1, x_2, \ldots, x_n) &= 0, \end{aligned}$$

where n is an integer, typically 2 or 3 for student problems. In optimisation the n different functions that are zero arise out of equating the n partial derivatives of a single function $f(x_1, x_2, \ldots, x_n)$ to zero. The basic idea of most numerical methods for solving such a system of equations is to make an informed guess and use properties of the functions to improve on this guess. If the guess is arbitrary, then one runs the risk of being so far away from the optimal solution required that no matter how sophisticated the routine, it is never reached. On the other hand, how do we make an informed guess? That is one question. Another is exactly what properties of the functions $f_k(x_1, x_2, \ldots, x_n)$, $k = 1, 2, \ldots, n$ does one use? These are the questions tackled in this case study. It should be mentioned here that if these functions are linear, then we are in the realm of linear programming. Linear programming is often taught within operational research which in turn is often thought of as a separate subject. "Mathematics, Statistics and Operational Research" is the title of some departments. In the United Kingdom, the QAA (Quality Assurance Agency) the national body that is designated as the guardian of the quality of UK higher education, also uses the title "Mathematics, Statistics and Operational Research" to cover mathematics and allied subjects. This title emphasises the separate nature of mathematics, statistics and operational research to some minds. Operational Research (or OR for short) really started life in the Second World War and has grown along with the ability of computers to solve systems with large arrays. The most useful problems involve maximising profit functions or minimising cost functions.

This is done by analysing the real problem and devising a number of inequalities obeyed by the variables. The commonest technique is the simplex method, and add-ons such as the use of hidden variables and integer programming are possible. This case study avoids linear programming, and concentrates on the cases where the functions are non-linear. It might be good to tackle a real problem involving specific choices of function that represent reality, on the other hand a more general approach means that the students can be introduced to extreme cases where pathological functions can be introduced to highlight the shortcomings of certain methods (and certain software!) The choice depends largely on student clientele and lecturer preference. Here, we shall be going down the more theoretical route as this case study is principally for mathematics undergraduates. Just in case there are students who have forgotten the Newton–Raphson method, it is a good idea to spend a few minutes on restating it. To solve the equation $f(x) = 0$ try a value x_0, then use the algorithm

$$x_k = x_{k-1} - \frac{f(x_{k-1})}{f'(x_{k-1})} \quad k = 1, 2, \ldots$$

to improve the guess. It is a good idea to draw the picture shown in Figure 5.5 which helps by showing several important features. First of all, if the first guess is poor, the slope of the function might make the next guess worse, or if the tangent to the curve is too close to the horizontal then $f'(x)$ is close to zero and the method fails. It is a good method once the first guess is appropriate. This gets the students in the right frame of mind for numerical work which is essentially practical. Do not go slavishly following a method which is taking ages to converge or is not working as it should. Try another "first guess" and start again! The Newton–Raphson method can be done for a cubic say or a function such as $f(x) = (x \sin x - 4 \cos x)$ and set up on MAPLE or even a spreadsheet such as Excel and animated as a class demonstration with a press of the return key getting a better and better approximation. The poor first guess can be similarly demonstrated. (A cubic with three real roots works well here–try to get the middle root!)

Moving on to functions of many variables, one obvious question is does the Newton–Raphson method generalise? The answer is yes, but it is not quite as useful. One of the most useful generalisations is the DFP (Davidon–Fletcher–Powell) method which we will meet later. First of all, as we have said optimi-

5. Case Studies

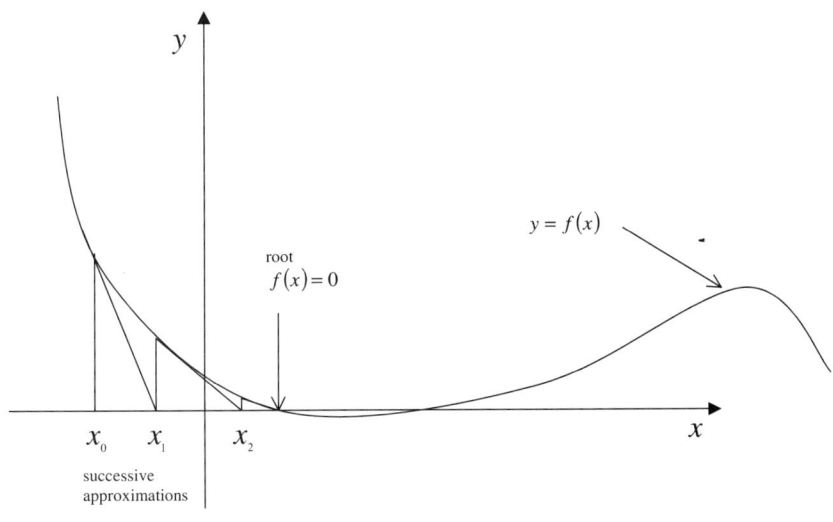

Figure 5.5   The Newton–Raphson technique.

sation involves solving the n partial derivatives of a function of n variables:

$$f_{x_1}(x_1, x_2, \ldots, x_n) = 0$$
$$f_{x_2}(x_1, x_2, \ldots, x_n) = 0$$
$$f_{x_3}(x_1, x_2, \ldots, x_n) = 0$$
$$\vdots$$
$$f_{x_n}(x_1, x_2, \ldots, x_n) = 0$$

where the notation:

$$f_{x_k} = \frac{\partial f}{\partial x_k}, \quad k = 1, 2, \ldots, n$$

has been adopted for convenience. It is neater to use vector notation as follows: write

$$\mathbf{x} = (x_1, x_2, \ldots, x_n)^T$$
$$\boldsymbol{\alpha} = (\alpha_1, \alpha_2, \ldots, \alpha_n)$$
$$\delta\boldsymbol{\alpha} = (\delta\alpha_1, \delta\alpha_2, \ldots, \delta\alpha_n)$$
$$\nabla = \left(\frac{\partial f}{\partial x_1}, \frac{\partial f}{\partial x_2}, \ldots, \frac{\partial f}{\partial x_n}\right)\Big|_{\mathbf{x}=\boldsymbol{\alpha}},$$

and the Hessian matrix of second order partial derivatives is defined as

$$\mathbf{G} = \begin{pmatrix} \frac{\partial^2 f}{\partial x_1^2} & \frac{\partial^2 f}{\partial x_1 \partial x_2} & \cdots & \frac{\partial^2 f}{\partial x_1 \partial x_n} \\ \frac{\partial^2 f}{\partial x_2 \partial x_1} & \frac{\partial^2 f}{\partial x_2^2} & \cdots & \frac{\partial^2 f}{\partial x_2 \partial x_n} \\ \vdots & \vdots & \ddots & \vdots \\ \frac{\partial^2 f}{\partial x_n \partial x_1} & \frac{\partial^2 f}{\partial x_n \partial x_2} & \cdots & \frac{\partial^2 f}{\partial x_n^2} \end{pmatrix} \bigg|_{\mathbf{x}=\boldsymbol{\alpha}}.$$

The first two terms of Taylor's theorem in n variables is now written as

$$f(\mathbf{x}) \approx f(\boldsymbol{\alpha}) + \delta\boldsymbol{\alpha}^T \boldsymbol{\nabla} f(\boldsymbol{\alpha}) + \frac{1}{2} \delta\boldsymbol{\alpha}^T \mathbf{G} \delta\boldsymbol{\alpha},$$

which approximates the function $f(\mathbf{x})$ near the point $\boldsymbol{\alpha}$. The direct analogy to the Newton–Raphson method for this n variable function would be the iterative scheme

$$\mathbf{x}_k = \mathbf{x}_{k-1} - [\mathbf{G}(\mathbf{x}_{k-1})]^{-1} \boldsymbol{\nabla} f(\mathbf{x}_{k-1}).$$

The problem is having to invert the Hessian matrix $\mathbf{G}(\mathbf{x})_{k-1}$ at each step which although theoretically possible makes for a very time consuming and lengthy process. It is also necessary to calculate second order derivatives. Are such complications really necessary? It is a good idea to discuss what the real problem is with the students before plunging into involved calculations. The gradient of the function $\boldsymbol{\nabla} f(\mathbf{x})$ gives a direction of steepest ascent and descent in n dimensional space. An analogy with walking over hills and down dales in two dimensions is always useful here. Therefore it might be possible to use this fact to home in on maxima or minima. Such methods, called methods of steepest descent, do not involve inverting matrices, and only the first derivatives of $f(\mathbf{x})$ need to be found. Students will ask what's the catch! The "catch" is that the method can be very slow. Although the calculations are more straightforward, there may be a lot of them. The basic iterative scheme is

1. choose $\mathbf{a}_k$

2. calculate $f(\mathbf{a}_k)$

3. calculate $\boldsymbol{\nabla} f$ and evaluate at $\mathbf{x} = \mathbf{a}_k$. Call this $\mathbf{h}_k$.

4. form $\mathbf{x} = \mathbf{a}_k - \lambda \mathbf{h}_k$

5. find the minimum value of the single variable λ, where $F(\lambda) = f(\mathbf{a}_k - \lambda \mathbf{h}_k)$.

From straightforward use of the chain rule, we have

$$\frac{dF}{d\lambda} = \sum_i \frac{\partial f}{\partial x_i} \frac{dx_i}{d\lambda} = -\boldsymbol{\nabla} f . \mathbf{h}_k.$$

This can now be used to determine the minimum value of λ, and hence determine the next approximation through

$$\mathbf{x}_{k+1} = \mathbf{x}_k - \lambda_{min}\mathbf{h}_k,$$

where $\mathbf{x}_k$ on the right is the initial guess $\mathbf{a}_k$. One possible student assignment is to send the students off to investigate such methods, possibly using a specific example such as Rosenbrock's function $10(x_1^2 - x_2)^2 - (1 - x_1)^2$ starting at the origin. Another is to investigate other optimisation methods such as the use of line searches and using conjugate gradients. Perhaps the most successful method of optimisation combines the simplicity of the steepest descent method with the rigour of the Newton–Raphson method. One method that does this is the Davidon–Fletcher–Powell method. Students need to realise that one is approximating already in the Newton–Raphson method through truncating the Taylor series, so further approximations brought in by inexact ways of inverting the Hessian matrix are acceptable as long as they are carefully done. Instead of working with $\mathbf{G}$ we define

$$\mathbf{x} = \mathbf{x}^{(r)} + t\mathbf{s}^{(r)}$$

where t is a parameter. This is a line in n space. Define

$$\mathbf{H}^{(r+1)} = \mathbf{H}^{(r)} + \mathbf{E}^{(r)}$$

where $\mathbf{E}^{(r)}$ is a correction term. The DFP algorithm (kth step) is as follows:

1. The search direction $\mathbf{p}^{(r)}$ is determined by $\mathbf{p}^{(r)} = -\mathbf{H}^{(r)}\mathbf{g}^{(r)}$ where $\mathbf{H}^{(r)}$ is the approximation to the inverse Hessian matrix $\mathbf{G}^{-1}$ and

$$\mathbf{g}^{(r)} = \boldsymbol{\nabla} f(\mathbf{x}^{(r+1)}) - \boldsymbol{\nabla} f(\mathbf{x}^{(r)}).$$

2. The step-length $t^{(r)}$ is determined by minimising on the lines given above using steepest descent, that is $t^{(r)}$ is the value of t that minimises the function $f(\mathbf{x}^{(r)} + t\mathbf{p}^{(r)})$.

3. Now update $\mathbf{x}^{(r)}$ using

$$\mathbf{x}^{(r+1)} = \mathbf{x}^{(r)} + t^{(r)}\mathbf{p}^{(r)}.$$

4. Finally, the next approximation to the inverse of the Hessian is given by

$$\mathbf{H}^{(r+1)} = \mathbf{H}^{(r)} - \frac{1}{\mathbf{y}^{(r)T}\mathbf{H}^{(r)}\mathbf{y}^{(r)}}\mathbf{H}^{(r)}\mathbf{y}^{(r)}\mathbf{y}^{(r)T}\mathbf{H}^{(r)} + \frac{1}{\mathbf{g}^{(r)T}\mathbf{p}^{(r)}}t^{(r)}\mathbf{p}^{(r)}\mathbf{p}^{(r)T}$$

where $\mathbf{y}^{(r)} = \mathbf{g}^{(r+1)} - \mathbf{g}^{(r)}$.

This algorithm looks complicated, and it only comes alive when examples are done. Here is the DFP algorithm used to try and find the minimum of the function
$$f(x_1, x_2) = x_1^2 - x_1 x_2 + 3x_2^2$$
The start vertex is $(1,1)$, that is $x_1 = 1$ and $x_2 = 1$. This gives $f(1,1) = 3$ and $\nabla f = (2x_1 - x_2, -x_1 + 6x_2)^T$. At the point $(1,1)$, $\nabla f = (1,5)^T$. The first step is thus
$$\mathbf{x} = \begin{pmatrix} 1 \\ 1 \end{pmatrix} + t \begin{pmatrix} 1 & 0 \\ 0 & 1 \end{pmatrix} \begin{pmatrix} 1 \\ 5 \end{pmatrix}$$
so
$$\begin{aligned} x_{11} &= 1 + t \\ x_{12} &= 1 + 5t. \end{aligned}$$
The parameter t is determined by minimising $f(1+t, 1+5t)$, that is
$$(1+t)^2 - (1+t)(1+5t) + 3(1+5t)^2.$$
Using standard methods, the answer is $t = -0.167$. The new value of $\mathbf{x}$ is thus
$$\mathbf{x_1} = \begin{pmatrix} 0.833 \\ -4.843 \end{pmatrix} \quad \text{with} \quad \nabla f_1 = \begin{pmatrix} 1.501 \\ 0.157 \end{pmatrix}.$$
The updated approximation to $\mathbf{H}$ is $\mathbf{H}_{k+1}$ and is given by the rather involved looking formula in the DFP algorithm. First we note that
$$\mathbf{h_0} = \mathbf{x_1} - \mathbf{x_0} = \begin{pmatrix} -0.167 \\ 0.157 \end{pmatrix}$$
and
$$\nabla f_1 - \nabla f_0 = \begin{pmatrix} 0.501 \\ -4.843 \end{pmatrix}$$
and so
$$\begin{aligned} \mathbf{H}_1 &= \begin{pmatrix} 1 & 0 \\ 0 & 1 \end{pmatrix} - \frac{1}{23.706} \begin{pmatrix} 0.251 & -2.426 \\ -2.426 & 23.455 \end{pmatrix} \\ &+ \frac{1}{3.96} \begin{pmatrix} 0.0279 & 0.139 \\ 0.139 & 0.697 \end{pmatrix} \\ &= \begin{pmatrix} 0.996 & 0.137 \\ 0.137 & 0.187 \end{pmatrix} \\ \mathbf{H}_1 \nabla f_1 &= \begin{pmatrix} 0.996 & 0.137 \\ 0.137 & 0.996 \end{pmatrix} \begin{pmatrix} 1.501 \\ 0.157 \end{pmatrix} = \begin{pmatrix} 1.517 \\ 0.235 \end{pmatrix}. \end{aligned}$$

The next step is to consider $f(0.833 - t1.517, 0.165 - t0.235)$ in order to find a value of t for which f is a minimum. This value is $t = 0.548$. To calculate the next value of $\mathbf{x}$ we use the formula

$$\mathbf{x}_2 = \mathbf{x}_1 - 0.548 \begin{pmatrix} 1.517 \\ 0.235 \end{pmatrix}$$

which equals $\mathbf{0}$ to a good approximation. The DFP algorithm has thus found the minimum in just two steps. This has to be the case for quadratic functions, and students should be asked why.

Further exercises for students could be to use software to solve more complicated problems. They could try using the DFP algorithm on "show" functions such as

$$(x_1 + 10x_2)^2 - 5(x_3 - x_4)^2 + (x_2 - 2x_3)^4 + 10(x_1 - x_4)^4$$

with start value $(3, -1, 0, 1)$, or

$$100(x_2 - x_1^2)^2 + (1 - x_1)^2 + 90(x_4 - x_3^2)^2 \ + \ (1 - x_3)^2 + 10.1((x_2 - 1)^2 \\ +(x_4 - 1)^2) \ + \ 19.8(x_2 - 1)(x_4 - 1)$$

with start value $(-3. - 1, -3, -1)$.

On both of these, using the Newton–Raphson method does not work and steepest descent is extremely slow.

The field of non-linear optimisation is only thirty or so years old and has burgeoned on the back of the exponential increase in computer power. It is a very fertile area for both theoretical and practical case studies, group projects and individual projects.

By contrast, this last case study reaches back into the History of Mathematics and should be accessible to all third year students without the use of a computer.

5.7 Euler and Series

The only prerequisites for this case study are knowledge of series, some trigonometry and elementary complex numbers, all normally found in mathematics curricula in the last years of school. They would certainly have been covered by the end of the first year at university.

Leonhard Euler had one of the finest mathematical minds of any era. This is not a controversial statement. The man left his mark on virtually all areas of mathematics, both pure and applied. If combinatorics is labelled as a rather pure area of statistics, then he can be said to have contributed to statistics too. He was born in 1707 in Switzerland, near Basel and spent the bulk of his professional life as Professor of Mathematics in St. Petersburg in Russia with a 25 year interlude (1741–1766) in Berlin. He died suddenly on 18th September 1783 aged 76, but not before writing more papers on mathematics than anyone else before or since. For more on this remarkably productive man see books on the History of Mathematics; I recommend especially *Euler, Master of us All* by William Dunham, published by the Mathematical Association of America in 1999 (185pp) which mingles biography with mathematical detail in a highly successful and entertaining way.

This case study focuses mainly on Euler's contribution to summing certain infinite series. It also touches on his pioneering work on complex numbers. Infinite series were a particular favourite with Euler, and he achieved much with them. Notice my reluctance to use the word 'proof'. Euler possessed an incredible intuition for arriving at correct results and he often used mathematically illegal shortcuts to get there. In the eighteenth century, the principles of mathematical rigour had not been firmly established. Advances in analysis occurred through Lagrange right at the end of the eighteenth century and in the next century through the work of Cauchy, Riemann, Gauss and others. This having been said, when Euler achieved a result he was aware of imperfections in his proof and often improved on his initial mathematics and re-derived results in many different ways. He had the uncanny knack of almost guessing correct results before establishing them with more and more rigour. He was the very essence of intuition and tenacity. For example, using the formula

$$\sin(A+B) = \sin A \cos B + \sin B \cos A$$

where A and/or B are complex numbers is completely unjustified as this formula is established through geometry in which both A and B are angles. However, using it Euler established de Moivre's theorem

$$(\cos\theta + i\sin\theta)^n = \cos n\theta + i\sin n\theta.$$

He went on to derive

$$e^{i\theta} = \cos\theta + i\sin\theta$$

from which comes the remarkable

$$e^{i\pi} + 1 = 0$$

an equation that links the five most important numbers in mathematics. All of this derived from what now would be considered unjustified mathematics. Through this kind of historical material, students are made aware of the changes in mathematical standards of rigour as well as developments in mathematical notation. On this latter point, all mathematics students should see how algebraic equations (linear, quadratic and cubic) were described in the fourteenth

5. Case Studies

and fifteenth century before the onset of modern notation. One then begins to understand how much harder mathematics can be with poor notation, and how much easier it might be with good notation. To return to Euler, one result worthy of study is his way (or ways) of finding the sum

$$\sum_{n=1}^{\infty} \frac{1}{n^2}.$$

Let us run through his first method. First of all the "infinite polynomial" (as Euler calls it) is defined by

$$P(x) = 1 - \frac{x^2}{3!} + \frac{x^4}{5!} - \frac{x^6}{7!} + \frac{x^8}{9!} + \cdots,$$

which sums to

$$P(x) = \frac{\sin x}{x}.$$

This result was reasonably well known in Euler's time. Now, $\sin x$ is zero wherever $x = n\pi$, $n = 0, \pm 1, \pm 2, \ldots$. Given this, Euler then reasons that

$$\sin x = x(1-\frac{x}{\pi})(1-\frac{x}{2\pi})\ldots(1+\frac{x}{\pi})(1+\frac{x}{2\pi})\ldots$$
$$= x(1-\frac{x^2}{\pi^2})(1-\frac{x^2}{4\pi^2})(1-\frac{x^2}{9\pi^2})\ldots.$$

Thus

$$\frac{\sin x}{x} = (1-\frac{x^2}{\pi^2})(1-\frac{x^2}{4\pi^2})(1-\frac{x^2}{9\pi^2})\ldots.$$

Questions of the convergence of the right hand side are left unanswered at this stage, but could be a discussion point with the class. Euler now expands both sides in powers of x, rearranging to give:

$$1 - \frac{x^2}{3!} + \frac{x^4}{5!} - \frac{x^6}{7!} + \frac{x^8}{9!} + \cdots = 1 - (\frac{1}{\pi^2} + \frac{1}{4\pi^2} + \frac{1}{9\pi^2} + \frac{1}{16\pi^2} + \frac{1}{25\pi^2} + \ldots)x^2 + \ldots.$$

Is the rearrangement of the infinite series on the right legal? Equating coefficients of x^2 yields

$$\frac{1}{3!} = \frac{1}{\pi^2}\left(1 + \frac{1}{4} + \frac{1}{9} + \frac{1}{16} + \frac{1}{25} + \cdots\right)$$

from which

$$1 + \frac{1}{4} + \frac{1}{9} + \frac{1}{16} + \frac{1}{25} + \ldots = \frac{\pi^2}{6}.$$

Euler calculated the left hand side numerically. Full frontal attacks are not useful as the series on the left converges only slowly. For example, summing the first 1000 terms gives an answer accurate to only two decimal places. Students should be asked to verify this, these days using a spreadsheet perhaps or a favourite piece of mathematical software. Euler had no such luxuries and used the following ingenious calculus trick. The improper integral

$$I = \int_0^{1/2} -\frac{\ln(1-t)}{t} dt$$

is evaluated in two different ways. Expanding the logarithm in the numerator and integrating term by term yields

$$\begin{aligned}
I &= \int_0^{1/2} -\frac{1}{t}\left(-t - \frac{t^2}{2} - \frac{t^3}{3} - \frac{t^4}{4} - \frac{t^5}{5} - \ldots\right) dt \\
&= \int_0^{1/2} \left(1 + \frac{t}{2} + \frac{t^2}{3} + \frac{t^3}{4} + \frac{t^4}{5} + \frac{t^5}{6} + \ldots\right) dt \\
&= t + \frac{t^2}{4} + \frac{t^3}{9} + \frac{t^4}{16} + \frac{t^5}{25} + \frac{t^6}{36} + \ldots \Big|_0^{1/2} \\
&= \frac{1}{2} + \frac{(1/2)^2}{4} + \frac{(1/2)^3}{9} + \frac{(1/2)^4}{16} + \frac{(1/2)^5}{25} + \frac{(1/2)^6}{36} + \ldots
\end{aligned}$$

Alternatively, substituting $x = 1 - t$ gives

$$\begin{aligned}
I &= \int_0^{1/2} -\frac{\ln(1-t)}{t} dt = \int_1^{1/2} \frac{\ln x}{(1-x)} dx \\
&= \int_1^{1/2} (1 + x^2 + x^3 + x^4 + \ldots) \ln x \, dx \\
&= \int_1^{1/2} \ln x \, dx + \int_1^{1/2} x \ln x \, dx + \int_1^{1/2} x^2 \ln x \, dx + \int_1^{1/2} x^3 \ln x \, dx + \ldots
\end{aligned}$$

The general result

$$\int_1^{1/2} x^n \ln x \, dx = \frac{x^{n+1}}{n+1} \ln x - \frac{x^{n+1}}{(n+1)^2} \Big|_1^{1/2}$$

can be used to evaluate the right hand side for $n = 1, 2, \ldots$ to give

$$\begin{aligned}
I &= (x \ln x - x) + \left(\frac{x^2}{2} \ln x - \frac{x^2}{4}\right) + \left(\frac{x^3}{3} \ln x - \frac{x^3}{9}\right) \\
&+ \left(\frac{x^4}{4} \ln x - \frac{x^4}{16}\right) + \ldots \Big|_1^{1/2} \\
&= \ln x \left(x + \frac{x^2}{2} + \frac{x^3}{3} + \frac{x^4}{16} + \ldots\right) - \left(x + \frac{x^2}{4} + \frac{x^3}{9} + \frac{x^4}{16} + \ldots\right) \Big|_1^{1/2} \\
&= \left[\ln\left(\frac{1}{2}\right)\right]^2 - \left(\frac{1}{2} + \frac{1/2^2}{4} + \frac{1/2^3}{9} + \frac{1/2^4}{16} + \ldots\right) + [\ln(0)\ln(1)] \\
&+ \sum_{n=1}^\infty \frac{1}{n^2}.
\end{aligned}$$

Thus, after convincing himself that $[\ln(0)\ln(1)]$ is zero(!) (an exercise in L'Hôpital's Rule for today's students) Euler arrived at

$$\sum_{n=1}^{\infty} \frac{1}{n^2} = 2\left(\frac{1}{2} + \frac{1/2^2}{4} + \frac{1/2^3}{9} + \frac{1/2^4}{16} + \ldots\right) + [\ln 2]^2$$

or

$$\sum_{n=1}^{\infty} \frac{1}{n^2} = \sum_{n=1}^{\infty} \frac{1}{n^2 2^{n-1}} + [\ln 2]^2$$

a sum which converges much more rapidly due to the power of two in the denominator of the right hand side sum. Using only fourteen terms, Euler obtained the answer

$$\sum_{n=1}^{\infty} \frac{1}{n^2} \simeq 1.644934$$

an answer we now know to be accurate to six decimal places. Evaluating $\pi^2/6$ verifies this nicely. Euler had got the right answer, but could all the mathematical steps be verified? Students could be asked to establish Wallis' formula

$$\frac{2}{\pi} = \frac{1.3.3.5.5.7.7.9.9\ldots}{2.2.4.4.6.6.8.8\ldots}$$

by putting $x = \pi/2$ in the $P(x)$ defined above. Another extension is to find

$$\sum_{n=1}^{\infty} \frac{1}{n^4}$$

and perhaps

$$\sum_{n=1}^{\infty} \frac{1}{n^6}$$

both of which Euler found, but which today can be obtained more easily. Investigations on the more general problem, finding

$$\sum_{n=1}^{\infty} \frac{1}{n^k}, \text{ where } k = \text{ an integer,}$$

leads students towards the realms of the Riemann zeta function and outside undergraduate mathematics.

Another fruitful extension, or more accurately parallel investigation, is the origin of the Euler constant $\gamma \simeq 0.577215\ldots$ a transcendental number defined by

$$\gamma = \lim_{n \to \infty} \left[\sum_{r=1}^{n} \frac{1}{r} - \ln(n+1)\right].$$

The results:
$$\gamma = -\int_0^\infty e^{-x} \ln x\, dx$$
and
$$\gamma = \lim_{x \to 1^+} \sum_{n=1}^\infty \left(\frac{1}{n^x} - \frac{1}{x^n} \right),$$
particularly the last, have a particularly pleasant symmetry and provide challenges for students.

5.8 Summary

In this chapter several case studies have been presented. The material could form a standard taught module: on the other hand each could be the subject of a group project or individual project. The distinctive features of case studies are first that there is a lot of presented material which means that the lecturer has more control over the direction the case study takes. In turn this usually means that the mathematics within the case study can afford to be more technical which opens up a wider class of problems for the student body. For example, it would be a very good student indeed who could get as far as the surface ocean dynamics case study outlined in section 5.2 through an individual project. Case studies thus give an opportunity to explore a specific piece of mathematics in a way such that the scope is very well defined by the lecturer. However, precisely because the lecturer is closely defining the scope, the mathematics can be more technical and the topic can be pushed further than is normal in an individual project.

Case studies can be assessed in a number of ways. Perhaps the most straightforward is to set coursework, but this has the usual danger that the submitted work is a group effort, or more unacceptably the work of one bright student who has allowed others to copy. An alternative is to set questions to be done under examination conditions. Longer than normal questions can be set, and the students could be allowed to take material into the examination room (open book examination). The questions would test the knowledge each student has of the case study while inviting him or her through the questions to take it a little further. Here are some examples of examination questions. Try some of them.

5.9 Exercises

1. By considering the equations for geostrophic flow and ignoring all vertical motion, show that the flow around a low pressure centre is anticlockwise in the northern hemisphere and clockwise in the southern hemisphere. If friction is included in the equations, deduce how this modifies the flow.

2. Solve the time dependent Ekman equations where they take the form
$$\frac{\partial W}{\partial t} + ifW = \nu_v \frac{\partial^2 W}{\partial z^2}$$
using Laplace Transform techniques. Use the following boundary conditions:
$$\frac{\partial W}{\partial z}(0,t) = \frac{1}{\rho\nu}\tau,$$
$$W \to 0 \quad \text{as} \quad z \to -\infty$$
and the sea is at rest initially. The complex form of the equations has been used whereby the horizontal flow (u,v) and the wind stress (τ^x, τ^y) are expressed as $W = u + iv$ and $\tau = \tau^x + i\tau^y$ with $i = \sqrt{-1}$. Do not attempt to invert the Laplace Transform, but use the final value theorem
$$\lim_{t \to \infty} W(z,t) = \lim_{s \to 0} s\bar{W}(s)$$
where s is the Laplace Transform variable to deduce both the functional form of the steady Ekman spiral, and the spin up time of this spiral.

3. Consider the damped Duffing equation
$$\ddot{x} + k\dot{x} + x - \frac{1}{6}x^3 = \Gamma \cos \omega t.$$
Follow the procedure in the notes by substituting $x = a(t)\cos\omega t + b(t)\sin\omega t$, assuming both $a(t)$ and $b(t)$ are slowly varying. Neglect $k\dot{a}$ and $k\dot{b}$, and hence show that to this degree of approximation
$$\dot{a} = -\frac{b}{2\omega}\{\omega^2 - 1 + \frac{1}{8}(a^2 + b^2)\} - \frac{1}{2}ka$$
$$\dot{b} = \frac{a}{2\omega}\{\omega^2 - 1 + \frac{1}{8}(a^2 + b^2)\} - \frac{1}{2}kb + \frac{\Gamma}{2\omega}.$$
Find the linear approximation in the neighbourhood of the equilibrium point when $\omega^2 > 1$. (Adapted from Jordan and Smith (1993).)

4. A pendulum with a light, rigid suspension is placed upside down on end, and the point of suspension is caused to oscillate vertically with displacement y given by $y = \epsilon \cos \omega t$, $\epsilon \ll 1$. Show that the equation of motion is

$$\ddot{\theta} + \left(-\frac{g}{a} + \frac{1}{a}\ddot{y}\right)\sin\theta = 0$$

where a is the length of the suspension, g is gravitational acceleration and θ the inclination to the vertical. Linearise this equation for small amplitude oscillations and show that Mathieu's equation is obtained. Investigate the equilibrium of this oscillation. (You may assume that Mathieu's equation

$$\ddot{x} + (\alpha + \beta \cos t)x = 0$$

has stability regions as given in Figure 5.6.)

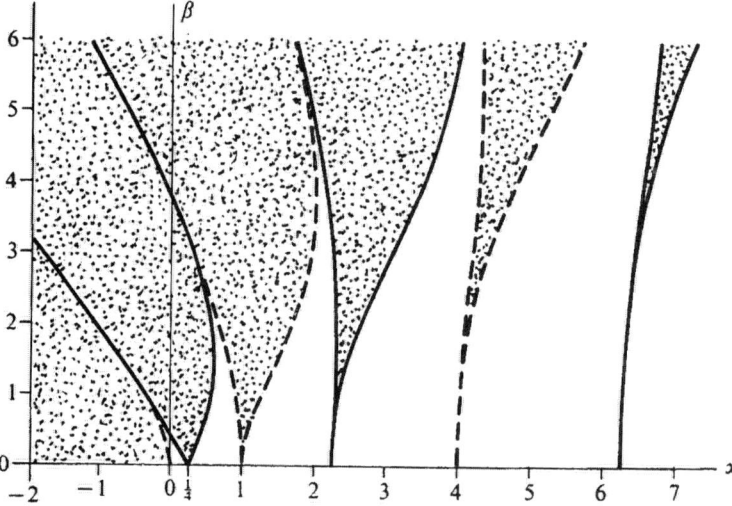

Figure 5.6 The stability diagram for Mathieu's equation. The shaded area is an unstable region in which at least one solution is unbounded, elsewhere all solutions are bounded. (Dashed line, solution period 2π exists; continuous line, solution period 4π exists). From *Nonlinear Differential Equations* by D.W.Jordan and P.Smith (Second edition, 1987). Reprinted by permission of Oxford University Press.

5. The traffic along a single lane is modelled using the equation

$$\frac{\partial \rho}{\partial t} + \frac{\partial}{\partial t}(\rho v(\rho)) = 0$$

where $\rho(x,t)$ is the density of cars and $v(\rho)$ their velocity. If

$$v(\rho) = v_0 \frac{\rho_{max} - \rho}{\rho_{max}}$$

and

$$\rho(x,0) = \rho_1 e^{-k|x|}$$

where k and $\rho_1 (\leq \rho_{max})$ are positive constants, determine when and where the solution first becomes undefined and hence a shock forms. Rework for the density given by

$$\rho(x,0) = \begin{cases} 0 & \text{for } x \leq 0 \\ \rho_1 x^2 / L^2 & \text{for } 0 \leq x \leq L \\ \rho_1 & \text{for } x \geq L \end{cases}$$

where L is a constant. (Adapted from Billingham and King (2000).)

6. The flow of cars along a single lane road can be described using a continuous car density $\rho(x,t)$ with a car velocity given by

$$v(\rho) = \frac{v_0}{\rho_{max}^2}(\rho_{max} - \rho)^2,$$

for $0 \leq \rho \leq \rho_{max}$. Determine the equation satisfied by ρ, and calculate the kinematic wave speed, $c(\rho)$. Show that $c(\rho)$ vanishes when $\rho = \frac{1}{3}\rho_{max}$ and has a minimum at $\rho = \frac{2}{3}\rho_{max}$.

At time $t = 0$ the car density is

$$\rho(x,0) = \frac{\rho_L + \rho_R e^{x/L}}{1 + e^{x/L}},$$

with $0 < \rho_R < \frac{1}{3}\rho_{max}$ and $\frac{2}{3}\rho_{max} < \rho_L < \rho_{max}$. Sketch the function $c(\rho(x,0))$. Also sketch the development of the car density for $t > 0$. (Adapted from Billingham and King (2000).)

7. Find a solution to the ordinary differential equation

$$w'' - 2zw' + 2kw = 0$$

of the form

$$\int_C e^{2z\zeta} f(\zeta) d\zeta$$

describing two possible contours C.

8. Prove that the differential equation

$$zw'' + 2aw' - zw = 0$$

where a is a real constant may be satisfied by assuming that

$$w = \int_C (t^2 - 1)^{a-1} e^{tz} dt,$$

where C is a suitable contour. Determine two allowable contours.

9. Determine the complete solution to the third order differential equation

$$zw''' + w = 0$$

by using contour integrals.

10. Investigate finding the minimum of the function

$$f(x_1, x_2, x_3) = 2(x_1 - 0.2)^4 + (x_2 - 0.1)^4 + 0.5(x_3 - 0.3)^4$$

using the method of steepest descent. Check your progress at each stage by comparing the approximation with the obvious correct solution.

11. Use the Davidon–Fletcher–Powell method to determine the minimum of the function

$$f(x, y, z) = (x - y)^2 + \frac{1}{16}(x + y + 1)^4$$

using two iterations. Show the the method of steepest descent is exact after only one iteration, but that the Newton–Raphson method converges only slowly. Give possible explanations.

12. Find at least two mathematically illegal steps in Euler's "proof" from considering the sine function that

$$\sum_{n=1}^{\infty} \frac{1}{n^2} = \frac{\pi^2}{6}.$$

Given that the result is correct, how are the steps Euler took in his proof validated?

13. "Euler's contribution to efficient mathematical notation is second to none." Discuss.

A
Project Example 1: Topics in Galois Theory

Preamble

For this project, the student explored an area of mathematics that was outside the course, and was difficult. The information was available and well embedded in the literature: however the concepts were advanced. In style therefore, this project most closely resembles the Hypergeometric Functions project outlined in Section 3.7.

Abstract

Galois Theory was originally formulated to determine whether the roots of a rational polynomial could be expressed in radicals. The theory is based on the association of a group to each polynomial and the analysis of the group to determine solubility in terms of radicals. We first study Galois Theory with the approach of the author Evariste Galois, then in its modern form based on abstract algebra, particularly field theory, which allows a more general interpretation. We then outline the theory of soluble groups and give some examples. Finally, we consider the application of methods in Galois Theory to the three classical construction problems and the construction of n-gons.

Introduction

Evariste Galois (1811–1832) was a French mathematician who was born near Paris. Although he died in tragic circumstances at 20 years old he had already produced work that would make him famous. Unfortunately during his lifetime Galois suffered many setbacks and was considered to be a troublemaker by the government. He also found himself unaccepted by the mathematical establishment. It was not until 1846 that Joseph Liouville published some of Galois' work in his *Journal de Mathematiques* [ED, p1]. Since then the importance of Galois' results has been realised and his work has provided the impetus for many developments in algebra.

In his *Memoir on the Conditions for Solvability of Equations by Radicals* Galois was solving a problem that interested many mathematicians including Lagrange. The solution to the general quadratic equation has been known since Babylonian times. In the 16th century formulae were discovered for the cubic and quartic. The solution for the cubic equation $x^3 + px = q$ found by Niccolo Fontana is

$$x = \sqrt[3]{\frac{q}{2} + \sqrt{\frac{p^3}{27} + \frac{q^2}{4}}} + \sqrt[3]{\frac{q}{2} - \sqrt{\frac{p^3}{27} + \frac{q^2}{4}}}.$$

This is an expression built up from the coefficients of the equation using the usual arithmetical operators with the nth roots, or radicals [ST, p xiv]. If the roots of an equation can be expressed in this way it is said to be "soluble by radicals". Mathematicians hoped that they would be able to find formulas for general equations of degree $n \geq 5$. Lagrange worked for many years on the solution of general equations, finding improved solutions for both the cubic and quartic [ED, pp18–22]. He was unable to find a solution for the quintic and expressed doubts that a solution existed. His work was almost certainly a source of inspiration for Galois. In 1824 Abel proved that the general quintic was not soluble by radicals so that a general formula did not exist but was unable to find a method for identifying which specific are soluble. This question was answered by Galois comprehensively in a piece of brilliant and original mathematics known as Galois Theory. This is the subject of this project and we now give an outline of each section.

In Section A.1 we closely follow the English translation of Galois' memoir by Harold M. Edwards in his book *Galois Theory* [ED]. The style, both of the translation and accompanying text, is traditional and often difficult to follow so much of the material in this section is a modified interpretation. The book takes a concrete rather than abstract approach to Galois Theory, and Edwards has attempted to present the theory with as little modern algebra as possible, for the most part drawing only on the techniques available to Galois. Some basic field theory is used in the text, particularly in the proofs, but we delay the introduction of fields until Section A.2.

In Section A.2 we study Galois Theory in its modern form which makes use of the developments in abstract algebra over the last 150 years. Field theory has allowed mathematicians to interpret Galois Theory in a formal, abstract sense, suited to the more rigorous standards of definition and proof introduced at the turn of the century. The main source for this chapter is Ian Stewart's *Galois Theory* [ST] which has a more accessible style than Edwards' book. In separating the two approaches to Galois Theory we are able to appreciate the simplicity of Galois' approach which the more structured methods of abstract algebra do not share although they allow Galois Theory to be applied to more general areas than rational polynomials.

Groups play an essential part in Galois Theory and Galois' work in this area provided a foundation for the study of group theory. This is expanded on in Section A.3, where we give an outline of the relevant theory and some examples. Sources for this chapter include Stewart's book and Allenby's *Rings, Fields and Groups* [AL] which was useful for a more detailed approach and worked examples.

Finally, Section A.4 explores the applications of field theory to classical geometry, using the methods developed in our study of Galois Theory. Here we combine elements from Stewart, Allenby and also John R. Durbin's *Modern Algebra: An Introduction*.

This project is intended to be an introduction to Galois Theory although, due to the depth of the subject and restrictions upon the project, it has not been possible to give more than an outline of many of the proofs. It is hoped however that it will provide the reader with the impetus for a more detailed study of the subject.

A.1 Galois' Approach

A.1.1 Preparation

We begin this section by reviewing some facts from basic algebra.

A.1.1.1 Polynomials. We consider polynomials with rational coefficients unless specifically declared otherwise. A polynomial of degree n is *irreducible* in Q if it cannot be expressed as a product of two or more polynomials whose degrees are less than n but greater than 0. The polynomial of degree 0 is the constant polynomial. The factorisation of a polynomial into irreducible polynomials is unique up to constant factors and the order in which the factors are written. If $f(x) = a_n x^n + \ldots + a_1 x + a_0$ and $a_n = 1$ then f is a *monic polynomial*.

Since we are interested in the roots of a polynomial, when two polynomials differ only by a constant factor we consider them to be essentially the same. This is demonstrated by the example

$$f(x) = \frac{1}{3}x^2 - \frac{1}{3}x - 2 = \frac{1}{3}(x+2)(x-3)$$
$$g(x) = x^2 - x - 6 = (x+2)(x-3).$$

So $f = \frac{1}{3} g$ but f and g have the same roots.

A.1.1.2 Some Methods of Checking Irreducibility. If f is a polynomial that is irreducible in $\mathbf{Z}$ then f is also irreducible in $\mathbf{Q}$. For example

$$\text{let} \quad g(x) = \frac{1}{3}x^2 + \frac{1}{3}x + 6$$
$$\text{and} \quad f(x) = x^2 + x + 18 = 3g(x).$$

Now f is irreducible in $\mathbf{Z}$ so f is irreducible in $\mathbf{Q}$.

Eisensteins's Irreducibility Criterion [ST, p20]

Let
$$f(x) == a_n x^n + \ldots + a_1 x + a_0, \; a_i \in \mathbf{Z}, \; i = 0, 1, \ldots, n.$$

Suppose there is a prime q such that $q \nmid a_n, q | a_i$ for $i = 0, 1, 2, \ldots, n-1$, and $q^2 \nmid a_0$ then f is irreducible in $\mathbf{Q}$.

For example, if $f(x) = x^5 + 4x^4 + 6x^3 + 2x^2 + 8x + 2$ and we take $q = 2$ then f is irreducible in $\mathbf{Q}$.

Unfortunately not every polynomial can be treated with this method. However, we can sometimes rewrite a polynomial so that the method will be suitable. For example, $f(x) = g(x)h(x)$ if and only if $f(x+1) = g(x+1)h(x+1)$. Hence $f(x)$ is irreducible if and only if $f(x+1)$ is irreducible [ST,p21].

An example of this is $f(x) = x^4 + x^3 + x^2 + x + 1$ where Eisenstein's criterion does not apply. But $f(x+1) = x^4 + 5x^3 + 10x^2 + 10x + 5$ and is irreducible by Eisenstein's criterion with $q = 5$. Hence $f(x)$ is irreducible in $\mathbf{Q}$.

A.1.1.3 Roots of Polynomials. Each root of a polynomial is either simple or multiple (repeated). We will want to work with polynomials that have only simple roots, but it is quite straightforward to detect simple roots.

If $f(x)$ has roots $\alpha_1, \alpha_2, \ldots, \alpha_n$ which are roots of multiplicity $m_1, m_2, \ldots, m_n$ respectively then

$$f(x) = (x - \alpha_1)^{m_1}(x - \alpha_2)^{m_2} \ldots (x - \alpha_n)^{m_n}.$$

If we differentiate formally then

$$Df = (x - \alpha_1)^{m_1-1}(x - \alpha_2)^{m_2-1}\ldots(x - \alpha_n)^{m_n-1} \sum_j m_j \prod_{i \neq j}(x - \alpha_i)$$

and it is clear that if, for any i, $m_i > 1$ then f and Df have a common factor. Thus we can say

Lemma 1 [ST,p85]

A polynomial has a multiple root if and only if f and Df have a common factor of degree ≥ 1. If f has no multiple roots then f and Df will be coprime, i.e. the highest common factor will be a polynomial of degree 0.

Any irreducible polynomial with rational coefficients must have only simple roots since if it had a multiple root then f and Df would have a common factor of degree≥ 1 which can only be a multiple of f. But f is irreducible and Df has degree less than n so we must have $Df = 0$ which implies that f is a constant.

A.1.1.4 Symmetric Polynomials. Symmetric polynomials play an essential role in Galois Theory as they allow us to express rational functions of the roots of a given equation in terms of the coefficients of the equation. Consider a general quadratic equation $f(x) = x^2 + bx + c$. If the roots of f are α and β then

$$\begin{aligned} f(x) &= (x - \alpha)(x - \beta) \\ &= x^2 - (\alpha + \beta)x + \alpha\beta. \end{aligned}$$

We can immediately see that $\alpha + \beta = -b$, $\alpha\beta = c$. The functions $\alpha + \beta$ and $\alpha\beta$ are called symmetric polynomials since their values do not vary if α and β are interchanged.

A *symmetric polynomial* is any function in n indeterminates that is invariant under any permutations of the indeterminates. For example, if the roots of a polynomial are $\alpha_1, \alpha_2, \alpha_3$ then the following functions are symmetric polynomials in the roots:

$$\alpha_1^2 + \alpha_2^2 + \alpha_3^2, \quad \alpha_1 + \alpha_2 + \alpha_3, \quad \alpha_1\alpha_2 + \alpha_1\alpha_3 + \alpha_2\alpha_3, \quad \alpha_1\alpha_2\alpha_3.$$

The last three examples are special cases of symmetric polynomials called *elementary symmetric polynomials*. It is these functions that are directly related to the coefficients of the polynomial of which the indeterminates are the roots.

Let $f(x) = x^n + b_1 x^{n-1} + b_2 x^{n-2} + \ldots + b_n = (x-r_1)(x-r_2)(x-r_3)\ldots(x-r_n)$. The elementary polynomials are

$$\begin{aligned}
\sigma_1 &= r_1 + r_2 + r_3 + \ldots + r_n = -b_1 \\
\sigma_2 &= r_1 r_2 + r_1 r_3 + \ldots + r_{n-1} r_n = b_2 \\
&\vdots \\
\sigma_n &= r_1 r_2 r_3 \ldots r_n = (-1)^n b_n
\end{aligned}$$

where $\sigma_j = 0$ for $j > n$. The general formula is $\sigma_k = (-1)^k b_k$.

Theorem 2 (Fundamental Theorem on Symmetric Polynomials) [ED, p9]

Every symmetric polynomial in $r_1, r_2, \ldots, r_n$ can be expressed as a polynomial in the elementary symmetric polynomials $\sigma_1, \sigma_2, \ldots, \sigma_n$. Moreover, a symmetric polynomial with integer coefficients can be expressed as a polynomial in $\sigma_1, \sigma_2, \ldots, \sigma_n$ with integer coefficients. (Proof [ED, pp9–12].)

Let $s_k = r_1^k + r_2^k + r_3^k + \ldots + r_n^k$ for $k = 1, 2, 3, \ldots$. Then we have the general formula [ED, p12]

$$s_k - s_{k-1}\sigma_1 + s_{k-2}\sigma_2 - \ldots + (-1)^{k-1} s_1 \sigma_{k-1} + (-1)^k k \sigma_k = 0 \quad k = 1, 2, 3, \ldots$$

We can also express the elementary symmetric polynomials in $n-1$ indeterminates in terms of elementary symmetric polynomials in n indeterminates. Let the indeterminates be $r_1, r_2, r_3, \ldots, r_n$ and let the elementary symmetric polynomials in the n indeterminates be σ_k. Let the elementary polynomials in $n-1$ indeterminates $r_1, r_2, r_3, \ldots, r_{n-1}$ be τ_k. Then [ED, pp10–11]

$$\begin{aligned}
r_1 &= \sigma_1 - r_n \\
r_2 &= \sigma_2 - r_n \sigma_1 + r_n^2 \\
r_3 &= \sigma_3 - r_n \sigma_2 + r_n^2 \sigma_1 - r_n^3 \\
&\vdots \\
r_n &= \sigma_{n-1} - r_n \sigma_{n-2} + \ldots + (-1)^{n-1} r_n^{n-1}
\end{aligned}$$

Finally, we note Lemma 1 from Galois' Memoir [ED, p51]:

Lemma 3 [ED, pp51–52]:

An irreducible equation cannot have a root in common with a rational equation without dividing it.

Proof: Let g and h be polynomials and g be irreducible and let g and h have a common root. Then using the Euclidean algorithm we can write $d = Ag + Bh$ where d is the greatest common divisor of g and h, and A and B are rational polynomials. If we substitute into this equation the common root, r, then we have $d(r) = 0$. Therefore d has degree > 0. Since d divides g and g is irreducible, d must be a non-zero multiple of g. Since d divides h it follows that g divides h.

□

A.1.2 The Galois Resolvent

A.1.2.1 Forming a Galois Resolvent. With the preliminary information detailed in Section A.1.1 we first study Galois Theory through Galois' original approach. The remainder of this chapter is drawn mostly from Edwards translation of Galois' Memoir [ED, Appendix 1, pp101–113] and accompanying text. As a starting point we have the "given" equation. Let this be

$$f(x) = x^n + a_{n-1}x^{n-1} + a_{n-2}x^{n-2} + \ldots + a_1x^1 + a_0.$$

This equation should be irreducible and have n distinct roots $\alpha_1, \alpha_2, \ldots, \alpha_n$. In addition, the coefficients of f are assumed to be rational.

Our aim is to show whether or not the roots of f can be expressed in terms of radicals. The only information available is the coefficients of the given equation. However, symmetric polynomials allow us to extract information about the roots from these coefficients.

Following earlier work by Lagrange ([AL, pp181–186] and [ED, pp18–22, pp32–35]) Galois constructed a function of the roots of the given equation. This function had particular properties and is now known as a Galois resolvent. The conditions governing the function are stipulated by Galois' Lemma II [ED, p102]:

Lemma 4: Given any equation with distinct roots $a, b, c \ldots$, one can always form a function V of the roots such that no two of the values one obtains by permuting the roots in this function are equal. Further, the coefficients of this function can be chosen as integers.

Proof:
Let our Galois resolvent be

$$V_0 = x_1\alpha_1 + x_2\alpha_2 + \ldots + x_n\alpha_n, \qquad x_i \in \mathbf{Z}.$$

We define a permutation of the roots by $\phi(\alpha_i) = \alpha_{\phi(i)}$. Any permutation $\phi \in S_n$ of the roots will give a new function of the roots which we will call a *conjugate* of the resolvent, say V_1. We shall denote the numerical value of any of the resolvent V_i by t_i. So if $\phi = (123\ldots n)$ then

$$\phi(V_0) = x_1\alpha_1 + x_2\alpha_2 + \ldots + x_n\alpha_n = V_1 \qquad V_0 \neq V_1$$

Note that we insist the *value* of each conjugate of the resolvent is different, not just the formal representation. We wish to show that the $n!$ conjugates have distinct values so let

$$D = \prod_{S,T} [x_1\left(S(\alpha_1) - T(\alpha_1)\right) + x_2\left(S(\alpha_2) - T(\alpha_2)\right) + \ldots]^2$$

where the product is over the $n!(n!-1)/2$ pairs of unordered permutations S and T in which $S \neq T$. If we regard the x_i as variables then D is a polynomial in the x_i with coefficients that are polynomials in the α_i. D is symmetric in the α_i so the coefficients are symmetric in the α_i. This means that D is a polynomial with known (rational) coefficients. Since D is a product of non-zero polynomials it is non-zero. We can therefore choose the x_i as integers so that $D \neq 0$.

□

A.1.2.2 Using a Galois Resolvent. Once we have a suitably chosen Galois resolvent we can use it in a series of calculations that yield information about the given equation f. We form the "equation for V" by constructing a polynomial with the resolvent and its conjugates as its roots. Let these be $V_0 = t_0, V_1 = t_1, \ldots, V_{n!-1} = t_{n!-1}$. Then the polynomial will be

$$F(X) = (X - t_0)(X - t_1)\ldots(X - x_{n!-1}).$$

Note that F is invariant under any permutation of the roots of f since this would only interchange the t_i. So the coefficients of F must be symmetric polynomials in the roots of f and are therefore known quantities. We can calculate these quantities as shown in the following example. This example will be used throughout sections A.1 and A.2 to demonstrate various aspects of Galois Theory.

Let $u(x) = x^3 - 2$. This is an irreducible polynomial with simple roots. Let the roots of u be a, b, c. We form the Galois resolvent, choosing coefficients that are integers. To make the calculations easier for this example, the coefficients have been chosen with the prior knowledge that they yield different values of the resolvent for each conjugation. Normally, such a calculation would need to be carried out without explicit values for the coefficient. Let the resolvent and its conjugates be

$$V_0 = a + 2b + 3c, \qquad V_1 = c + 2a + 3b, \qquad V_2 = b + 2c + 3a$$
$$V_3 = b + 2a + 3c, \qquad V_4 = c + 2b + 3a, \qquad V_5 = a + 2c + 3b.$$

A. Project Example 1: Topics in Galois Theory

Then the polynomial with the values of the resolvent and its conjugates as roots is

$$\begin{aligned}
U(X) &= (X - t_0)(X - t_1)(X - t_2)(X - t_3)(X - t_4)(X - t_5) \\
&= X^6 - (t_0 + t_1 + t_2 + t_3 + t_4 + t_5)X^5 \\
&\quad + [(t_0 + t_1)(t_2 + t_3 + t_4 + t_5) + (t_2 + t_3)(t_4 + t_5)] X^4 \\
&\quad + [t_0 t_1 + t_2 t_3 + t_4 t_5] X^4 \\
&\quad - [t_0 t_1(t_2 + t_3 + t_4 + t_5) + t_2 t_3(t_4 + t_5) + t_4 t_5(t_2 + t_3)] X^3 \\
&\quad + [(t_0 + t_1)(t_2 t_3 + t_4 t_5 + (t_2 + t_3)(t_4 + t_5))] X^3 \\
&\quad + [t_2 t_3 t_4 t_5 + (t_0 + t_1)(t_2 t_3(t_4 + t_5))] X^2 \\
&\quad + [(t_4 t_5(t_2 + t_3)) + t_0 t_1((t_2 + t_3)(t_4 + t_5) + t_2 t_3 + t_4 t_5)] X^2 \\
&\quad - [(t_0 + t_1)t_2 t_3 t_4 t_5 + t_0 t_1(t_2 t_3(t_4 + t_5) + t_4 t_5(t_2 + t_3))] X \\
&\quad + t_0 t_1 t_2 t_3 t_4 t_5.
\end{aligned}$$

If we expand the coefficients of U in terms of the roots a, b and c we find that they are symmetric polynomials in the roots. For example the terms for X^5 and X^4 are

$$12(a + b + c)X^5$$

$$\left[58(a^2 + b^2 + c^2) + 122(ab + ac + bc)\right] X^4.$$

Using the symmetric polynomials and elementary symmetric polynomials as defined in Section A.1 we have the following information:

$$\sigma_1 = \sigma_2 = 0, \qquad \sigma_3 = 2$$

$$s_1 = s_2 = s_4 = s_5 = 0, \qquad s_3 = 6, \qquad s_6 = 12.$$

So our polynomial becomes

$$\begin{aligned}
U(X) &= X^6 + t_0 t_1 t_2 t_3 t_4 t_5 \\
&= X^6 + 108
\end{aligned}$$

which is irreducible in **Q**.

In general the polynomial F will not always be irreducible but will decompose into irreducible factors so that $F(X) = G_1(X)G_2(X)G_3(X)\ldots G_s(X)$. Note that since the roots of F are the conjugates of the resolvent then each of the factors must have some of the conjugates as its roots, and each factor has roots distinct from the roots of each other factor. We choose G_1 to be the factor that has $V_0 = t_0$ as a root.

A.1.2.3 Expressing roots in terms of the Galois Resolvent. We now move on to Galois Lemma III [ED, p103]

Lemma 5: When the function V is chosen as indicated above, it will have the property that all the roots of the given equation can be expressed as rational functions of V.

Proof: First, Galois considered the conjugates of the resolvent that held one, arbitrarily chosen, root fixed and permuted all the others. There are $(n-1)!$ of these conjugates and we use them to construct a polynomial with them in the same manner we used for F. V is always one of these conjugates that holds a root fixed. Let the others be $V_1, V_2, \ldots$ then

$$F_a(X) = (X - t_0)(X - t_i)(X - t_{i+1}) \ldots (X - t_{i+s}).$$

F_a is symmetric in the roots that have been permuted. It can be shown that any one root of a given equation can be expressed in terms of elementary symmetric polynomials in all the other roots [ED, pp10–11]. So we can eliminate from F_a all but the fixed root, and we can say

$$F_a(X) = F_a(X, \alpha_1) \qquad \text{where } \alpha_1 \text{ is the fixed root.}$$

Replace α_1 with the indeterminate Y and substitute t_0 for X so that we have $F_a(t_0, Y)$ and $f(Y)$ as polynomials in Y with coefficients that are rational functions in t_0. We find the greatest common divisor $d(Y)$ of $F_a(t_0, Y)$ and $f(Y)$. Now $(Y - \alpha_1)$ divides $F_a(t_0, Y)$ and $f(Y)$ so that $d(Y) = A(Y)F_a(t_0, Y) + B(Y)f(Y)$. Setting $Y = 0$ gives

$$\gamma(-\alpha_1) = A(0)F_a(t_0, 0) + B(0)f(0)$$
$$\alpha_1 = h_1(t_0)$$

where h_1 is a polynomial in t_0 with rational coefficients. We can perform a similar process for each of the roots in turn. If we take the conjugates of the resolvent that hold another of the roots fixed we form another equation, say $\alpha_2 = h_2(t_0)$. This is most clearly demonstrated by returning to the example $u(x) = x^3 - 2$ with roots a, b, c. We see that V_0 and V_5 leave a fixed:

$$U_0 = (X - t_0)(X - t_5)$$
$$= X^2 - (t_0 + t_5)X + t_0 t_5.$$

Now $t_0 + t_5 = 2a + 5(b+c)$ and $t_0 t_5 = a^2 + 6(a^2 + b^2) + 5a(b+c) + 13bc$. If τ_i are the elementary symmetric polynomials in b and c then we have

$$\tau_1 = b + c = -a, \qquad \tau_2 = bc = a^2, \qquad b^2 + c^2 = \tau_1^2 - 2\tau_2 = -a^2.$$

A. Project Example 1: Topics in Galois Theory

So $U_a(X,a) = X^2 + 3aX + 3a^2$. The greatest common divisor of U_a and u is d where
$$d(Y) = \frac{1}{3}(2t_0^2 Y + t_0^3 - 6).$$

Setting $d(Y) = \gamma(Y - \alpha_1)$ and $Y = 0$ we find $\gamma = \dfrac{2t_0^2}{3}$ and $a = h_1(t_0) = \dfrac{6 - t_0^3}{2t_0^2}$ and similarly we can show $b = h_2(t_0) = -\dfrac{6}{t_0^2}$, $c = h_3(t_0) = \dfrac{6 + t_0^3}{2t_0^2}$.

□

Since the conjugates of the resolvent are obtained by a permutation of the roots of the given equation f we should expect that the roots can be expressed in terms of all of these conjugates. We have Galois' Lemma IV [ED, p104]:

Lemma 6:
Suppose one has formed the equation for V, and that one has taken one of its irreducible factors, so that V is the root of an irreducible equation. Let $V, V', V'', \ldots$ be the roots of this irreducible equation. If $a = f(V)$ is one of the roots of the given equation, $f(V')$ will also be a root of the given equation.

Proof:
In the notation we are using we have
$$f(X) = G_1(X) G_2(X) G_3(X) \ldots G_s(X)$$
where t_0 is a root of G_1 which has deg. G_1 roots taken from the t_i. So if we have $\alpha_1 = h_1(t_0)$ then perform the permutation $\theta = (123\ldots n)$ we have $\alpha_2 = h_1(\theta(t_0))$.

□

Returning to our example, we have $a = h_1(t_0) = \dfrac{6 - t_0^3}{2t_0^2}$. The permutation (132) takes t_0 to t_2 and a to c so $c = h_1(t_2) = \dfrac{6 - t_2^3}{2t_2^2}$.

A.1.3 The Galois Group

A.1.3.1 Finding the Galois Group. We come now to a fundamental part of Galois Theory. Any given equation has a particular group associated with it and in this section we show how Galois found this group. Firstly we quote Galois' Proposition 1 [ED, p104]:

Theorem 7:

Let an equation be given whose m roots are $a, b, c, \ldots$. There will always be a group of permutations of the letters $a, b, c, \ldots$ which will have the following property:

1. That each function invariant under the substitutions of this group will be known rationally,

2. Conversely, that every function of the roots which can be determined rationally will be invariant under these substitutions.

We have $F(X) = G_1(X)G_2(X)G_3(X)\ldots G_s(X)$. Let deg G_1 be m so the roots of G_1 are m of the t_i including t_0. The given equation f has n roots that can all be represented as functions of the t_i. We can construct a table as below:

$$h_1(t_0), \quad h_2(t_0), \quad h_3(t_0), \quad \ldots \quad h_n(t_0)$$
$$h_1(t_i), \quad h_2(t_i), \quad h_3(t_i), \quad \ldots \quad h_n(t_i)$$
$$\vdots \qquad \vdots \qquad \vdots \qquad \vdots \qquad \vdots$$
$$h_1(t_j), \quad h_2(t_j), \quad h_3(t_j), \quad \ldots \quad h_n(t_j).$$

This table has m rows representing the m conjugates of the resolvent and n columns representing the roots of the given equation f. This table gives us the Galois group since each row represents one arrangement of the roots. We can demonstrate this by using our example $u(x) = x^3 - 2$ once again. The table of functions of the t_i is

$$h_1(t_0), \quad h_2(t_0), \quad h_3(t_0)$$
$$h_1(t_1), \quad h_2(t_1), \quad h_3(t_1)$$
$$h_1(t_2), \quad h_2(t_2), \quad h_3(t_2)$$
$$h_1(t_3), \quad h_2(t_3), \quad h_3(t_3)$$
$$h_1(t_4), \quad h_2(t_4), \quad h_3(t_4)$$
$$h_1(t_5), \quad h_2(t_5), \quad h_3(t_5)$$

which gives

$$a \quad b \quad c$$
$$c \quad a \quad b$$
$$b \quad c \quad a$$
$$b \quad a \quad c$$
$$c \quad b \quad a$$
$$a \quad c \quad b.$$

These permutations are those contained in the whole of the symmetric group S_3. So the Galois group is for u in S_3.

Proof of Theorem 7:

This breaks down into several parts. We need to show:

A. Project Example 1: Topics in Galois Theory

i) the entries in the table are all roots of f,

ii) no root appears in a row more than once;

iii) points 1 and 2 regarding invariant and rational functions;

iv) closure of the group represented in the table.

i) $f(h_i(X))$ is a polynomial in X with rational coefficients. Now $h_1(t_0) = \alpha_1$ so $f(h_1(t_0)) = 0$. This shows that t_0 is a root of $f(h_1(X))$. But t_0 is also a root of G_1, an irreducible equation. By Lemma 3 we must have G_1 divides $f(h_1)$. So every root of G_1 is also a root of $f(h_1)$. Similarly, every root of G_1 is a root of $f(h_2), f(h_3)$, etc. So all the entries in the table are roots of f [ED, p52].

ii) Let α_m and α_n be two roots of f so that $h_m(t_0) = \alpha_m$ and $h_n(t_0) = \alpha_n$. Suppose

$$h_m(t_i) = h_n(t_i)$$
$$h_m(t_i) - h_n(t_i) = 0$$

where t_i is a root of G_1. Then t_i is a root of $h_m(X) - h_n(X)$. So by Lemma 3 G_1 divides $h_m(X) - h_n(X)$ and t_0 is a root also. But then $h_m(t_0) = h_n(t_0)$, $\alpha_m = \alpha_n$ and the roots of f are distinct. So we cannot have $h_m(t_i) = h_n(t_i)$ and the entries of each row of the table are distinct.

iii) Let $P(\alpha_1, \ldots, \alpha_n)$ be a rational polynomial in the n roots of f, such that P is invariant under any permutation of the roots. Since each root can be expressed as a function of the Galois resolvent t_0 we have

$$P(\alpha_1, \ldots, \alpha_n) = P(h_1(t_0), \ldots, h_n(t_0)).$$

But since P is invariant under the permutations of the roots we must have

$$P(h_1(t_0), \ldots, h_n(t_0)) = P(h_1(t_i), \ldots, h_n(t_i)) \quad \text{for all } t_i \text{ roots of } G_1.$$

Let $(h_1(t_i), \ldots, h_n(t_i)) = \underline{h}(t_i)$. So we can say

$$P(\underline{h}(t_0)) = \frac{1}{m}[P(\underline{h}(t_0)) + P(\underline{h}(t_i)) + \ldots] \quad \text{where } m = \deg G_1$$

But this is a symmetric polynomial in the roots of G_1 and can therefore be expressed in terms of the rational coefficients of G_1. So

$$P(\underline{h}(t_i)) \in \mathbf{Q} \Rightarrow P(\alpha_1, \ldots, \alpha_n) \in \mathbf{Q}.$$

Conversely, assume that $P(\alpha_1, \ldots, \alpha_n) \in \mathbf{Q}$ so that $P(\underline{h}(t_i)) \in \mathbf{Q}$. Now $P(\underline{h}(X))$ is a polynomial with rational coefficients so $P(\underline{h}(X)) - P(\underline{h}(t_0))$ is also a rational polynomial which has a root t_0. Therefore by Lemma 3 $P(\underline{h}(X)) - P(\underline{h}(t_0))$ is divisible by G_1 and has the roots of G_1 in common.

But then $P(\underline{h}(t_i)) = P(\underline{h}(t_0))$ for any t_i a root of G_1. So P is invariant under any permutation of the roots of f.

iv) To prove closure we show that no matter which row arrangement we begin with we obtain the same group of permutations. Let s and t be two different Galois resolvents for f, let ϕ be a permutation belonging to the Galois group of f and let α_1, α_2 be the roots of f such that $\alpha_2 = \phi(\alpha_1)$.

Suppose $\alpha_1 = h_s(s)$ and $\alpha_1 = h_t(t)$ then $\alpha_2 = \phi(h_s(s)) = h_s(\phi(s))$. We want to show that $\phi(h_t(t)) = \alpha_2$.

Now

$$\begin{aligned} h_s(s) &= h_t(t) \\ h_s(s) - h_t(t) &= 0 \quad 0 \in \mathbf{Q} \\ \phi(h_s(s) - h_t(t)) &= 0 \\ h_s(\phi(s)) &= h_t(\phi(t)) \\ \alpha_2 &= h_t(\phi(t)). \end{aligned}$$

Since the elements of the group are independent of the choice of Galois resolvent, they must also be independent of the choice of which conjugate of the resolvent we begin with.

$\square$

A.1.3.2 Galois' Examples of Groups. Galois highlights two examples of groups. The first is the general polynomial of degree n where the coefficients are not explicitly known but are rational. In this case the only functions that can be determined rationally are the symmetric polynomials in the roots. Galois states, without proof, that the Galois group of a general polynomial of degree n is the symmetric group S_n containing $n!$ permutations of the roots. We delay discussion of this point until Section A.2, accepting for the moment the statement without proof.

The second example given is the cyclotomic equation of degree n

$$\frac{x^n - 1}{x - 1} = x^{n-1} + x^{n-2} + \ldots + x + 1 \quad \text{where } n \text{ is prime.}$$

Galois shows that the group for this polynomial consists of n permutations only. We will use this group as an example in Section A.3 which concentrates on soluble groups.

A.1.4 Soluble Equations and Soluble Groups

A.1.4.1 Adjoining Quantities to an Equation. At this point it is worth reviewing our progress so far. For our given equation f we can find a Galois resolvent, say V_0, and form an equation F of degree $n!$ such that

$$\begin{aligned} F(X) &= \prod_{i=0}^{n!-1} (X - t_i) \\ &= G_1(X)G_2(X)G_3(X)\ldots G_s(X) \end{aligned}$$

where the G_i are irreducible polynomials with rational coefficients and roots t_i. We choose G_1 to be the polynomial with one of its roots t_0. By expressing each root of f as a function of t_0 we can construct a table which leads us to the Galois group of f. Importantly, the number of rows in the table, which is deg G_1, is the order of the Galois group.

We now move on to the analysis of the Galois group. Galois was interested in how the group changed as certain quantities were adjoined to the given equation f. If we wish to express the roots of f in radicals we will adjoin the required radicals, which will all be of smaller degree than n. In fact, we need only adjoin radicals of prime order or, if we want the jth root where j is not a prime, then we decompose j into the product of primes. Say $j = j_1 j_2 j_3 \ldots$ then we can construct a j_1th root of a j_2th root of a j_3th root, etc. So we are only interested in adjoining pth roots where p is a prime.

We are interested in adjoining sufficient quantities to the given equation f to allow us to decompose it into linear factors. If this is possible then the equation is obviously soluble by radicals. We have Galois' Proposition II [ED, p106]:

Theorem 8:
   If one adjoins to a given root r of an auxiliary irreducible equation,

1. Either the group of the equation will not be changed, or it will be partitioned into j groups where j divides p, each belonging to the given equation respectively when one adjoins each of the roots of the auxiliary equation; if p is prime then the partition is into 1 or p groups.

2. These groups will have the remarkable property that one will pass from one to the other in applying the same substitution of letters to all the permutations of the first.

The auxiliary irreducible equation that Galois had in mind is of the form $x^p = k$, the solution of which gives us primitive pth roots of k. Let these roots be $r, \alpha r, \alpha^2 r, \ldots, \alpha^{p-1} r$ where α is a pth root of unity. Gauss' work had shown that pth roots of unity are expressible in radicals for any prime p (although

Edwards disagrees [ED, p27]). So Galois always assumed that pth roots of unity had already been adjoined to the given equation.

Proof:

Galois does not give full proof, but Edwards' detailed proof [ED, pp59-61] is outlined here. If G_1 remains irreducible then the Galois group of the given equation f remains unchanged. Otherwise, G_1 reduces to the product of irreducible factors $G_1(X) = H_1(X)H_2(X)\ldots H_v(X)$. The new Galois group is of order deg H_1 and consists of the arrangements of the roots of f expressed as functions of those conjugates of V_0 that are roots of H_1. Again we must choose H_1 to be the factor of G_1 that has $V_0 = t_0$ as a root. Since we have adjoined to f the root r of $x^p = k$ we can represent H_1 as $H_1(X, r)$. Edwards shows that by substituting $\alpha^i r$ for $i = 1, \ldots, p-1$ into H_1 we can form the product

$$h(x) = H_1(X, r)H_1(X, \alpha r) \ldots H_1(X, \alpha^{p-1} r).$$

Since G_1 and H_1 have deg H_1 roots in common and coefficients that are rational functions of r then by Lemma 3 $G_1(X) = H_1(X, \alpha^i r)Q(X, r)$ where Q is a polynomial with coefficients that are also rational functions of r. We can also show that

$$G_1(X) = H_1(X, \alpha^i r)Q(X, \alpha^i r) \qquad \text{for } i = 1, \ldots, p-1.$$

So we can say $G_1(X)^p = h(X)q(X)$ and by Lemma 3

$$1 = \left[\frac{h(X)}{G_1(X)^j}\right]\left[\frac{q(X)}{G_1(X)^{p-j}}\right].$$

So $h(X) = \text{constant} \times G_1(X)^j$ for some integer j. This gives us deg $h = p$deg $H_1 = j$deg G_1. The index of the new Galois group in the old Galois group is deg G_1/deg H_1 and this, by the above equation, is p/j. If p is prime then either $j = 1$ or p so the index is 1 or p. Obviously, the value of 1 relates to G_1 remaining irreducible. To show the second part of Theorem 8 we first note that when j is 1 we have

$$\text{constant} \times G_1(X) = H_1(X, r)H_1(X, \alpha r) \ldots H_1(X, \alpha^{p-1} r).$$

The roots of $H_1(X, r)$ are a subset of the roots of $G(X)$, as are the roots of $H_1(X, \alpha r)$. Let the two roots of $H_1(X, r)$ be t_0 and t_1 and a root of $H_1(X, \alpha r)$ be t_2. A row of the group table for H_1 is

$$\alpha_1 = h_1(t_0), \alpha_2 = h_2(t_0), \ldots, \alpha_n = h_n(t_0).$$

If S is a permutation belonging to the new Galois group and $S(t_0) = t_2$ then

$$S(\alpha_1) = h_1(t_2), S(\alpha_2) = h_2(t_2), \ldots, S(\alpha_n) = h_n(t_2)$$

is a row of the table for $H_1(x, \alpha r)$. Edwards shows that if we apply S to the row of the table which contains functions of r_1 then we obtain $h_1(t_3), h_2(t_3), \ldots, h_n(t_3)$ where t_3 is also a root of $H_1(X, \alpha r)$. So by applying the same permutation we move from the new Galois group to another subset of the old Galois group, relating to another factor of G_1. A subgroup with this property is known as a *normal subgroup*. A normal subgroup partitions a group into conjugacy classes. In the above context these conjugacy classes are the subsets of the old Galois group relating to each irreducible factor H_1 of G_1. As a reminder, a subgroup B of a group A is normal if and only if for $a \in A$ and $b \in B$ we have $a^{-1}ba \in B$.

The process of adjoining quantities is continued until the Galois group of f has been reduced to the identity permutation only. At this stage we have adjoined all the necessary radicals to f and F has reduced to linear factors. Then the equation of f is soluble for its roots.

□

This shows that if f is soluble by radicals then its Galois group will reduce by a series of subgroups, each a normal subgroup of prime index in its predecessor. Let the Galois group of f be B_0. Ignoring any groups that are identical to their predecessors, we have the series

$$B_0 \triangleright B_1 \triangleright B_2 \triangleright \ldots \triangleright B_v = e \text{ where } |B_{i-1}|/|B_i| = p \text{ with } p \text{ prime and } i = 1, \ldots, v.$$

We call a group *soluble* if it has such a series of subgroups.

Consider our example $u(x) = x^3 - 2$. If we adjoin a primitive cube of unity, y, then decompose into two factors:

$$\begin{aligned} U(X) &= (X^3 - 2(75y^2 + 69y + 72))(X^3 - 2(69y^2 + 75y + 72)) \\ &= V_1(X)V_2(X). \end{aligned}$$

The irreducible factor V_1 has Galois group A_3 which is the alternating group of degree 3. A_3 is a normal subgroup of S_3 of index 2. If we now adjoin z, a real cube root of 2, then U can be expressed in linear factors since the roots of u are $a = z$, $b = yz$, and $c = y^2z$. This gives

$$\begin{aligned} U(X) =& (X - (z + 2yz + 3y^2z))(X - (y^2z + 2z + 3yz))(X - (yz + 2y^2z + 3z)) \\ & \times (X - (yz + 2z + 3y^2z))(X - (y^2z + 2yz + 3z))(X - (z + 2y^2z + 3yz)). \end{aligned}$$

So the series of subgroups for the Galois group of u is $S_3 \triangleright A_3 \triangleright e$ where e is a normal subgroup of A_3 with index 3.

Galois' Propositions III and IV [ED, p107] can be dealt with fairly briefly. Proposition III states:

Theorem 9:

If one adjoins to an equation all the roots of an auxiliary equation, the groups in Theorem II [Theorem 8 here] will have the further property that each group contains the same substitutions.

Our choice of r, as a solution of the auxiliary equation, is completely arbitrary so we must necessarily obtain the same partitioning of the group for each root $r, r', r^n, \ldots, r^{p-1}$ of the auxiliary equation. If we adjoin all these roots then rather than splitting the Galois group of f into conjugacy classes, each relating to one irreducible factor of G_1, we have the same normal subgroup of the Galois group applying to each factor.

Theorem 10:

If one adjoins to an equation the numerical value of a certain function of its roots, the group of the equation will be reduced in such a way as to contain no permutations other than those which leave this function fixed.

A.1.4.2 Finding Soluble Groups and Soluble Equations. Proposition V [ED, p108] asks the question which is the motivation behind Galois Theory–"In which case is an equation solvable by simple radicals?"

We know that the group of the equation must have a series of normal subgroups of index p which relate to the adjunction of pth roots and the series must end with the identity permutation (a fuller discussion of soluble groups follows in Section A.3). Galois gives a rather compact proof of this [ED, p109]. Edwards expands this proof considerably covering some points missed by Galois [ED, pp61–64].

Galois gives the example of the general quartic equation [ED, p109]. The group has 24 elements and the adjunction of a square root decomposes the group into two subsets of order 12. The new Galois group is A_4 which is a normal subgroup of S_4 of index 2. Then a cube root is adjoined and A_4 splits into 3 subsets. The new Galois group is $\{e, (12)(34), (13)(24), (14)(23)\}$ which is a normal subgroup of A_4 of index 3. We then adjoin another square root to give a Galois group $\{e, (12)(34)\}$ which is a normal subgroup of the previous group, of index 2. Finally we adjoin another square root to reduce the Galois group to the identity permutation. Therefore the general quartic is soluble by radicals.

More interesting are the general equations of degree ≥ 5. Consider the general quintic equation. The Galois group is S_5 so we need to find whether S_5 is soluble. Here we quote a theorem which Edwards has derived from Galois' memoir [ED, p91]

Theorem 11:

If an irreducible equation has prime degree p and is solvable by radicals, then the roots of the equation can be ordered $a_1, a_2, \ldots, a_p$ in such a way that the permutations s of the Galois group are all of the form $s(a_i) = a_{ri+s}$ where a_i is defined for all integers i by setting $a_i = a_j$ for $i \equiv j(\mathrm{mod} p)$, where r and s are integers, and where $r \neq 0(\mathrm{mod} p)$. The proof can be found in Edwards [ED, pp 91–93].

If the general quintic equation is soluble by radicals then, by the above theorem the symmetric group S_5 should contain only permutations of the form $s(a_i) = a_{ri+s}$. If we are working modulo 5 then there are 5 values of s but only 4 values of r that we can use. This gives 20 possible permutations but S_5 has 120 elements. Therefore the general quintic equation is not soluble by radicals [ED, p91]. This does not mean that every quintic is insoluble, but it is not possible to express the roots in terms of a formula in the coefficients of the equation using arithmetical operators and radicals.

Since S_5 is a subgroup of every symmetric group S_n for $n \geq 5$ no symmetric group is soluble for $n \geq 5$ [ED, p90]. This means that there is no formula for the solution of general equations of degree greater than the general quartic.

A.2 The Modern Approach

A.2.1 Field Extension

A.2.1.1 Constructing Field Extensions. For this section we shall assume that the reader understands the basics of ring, field and group theory although some theorems will be quoted (without proof) to highlight their importance. A useful basic text is Allenby's *Rings, Fields and Groups* [AL] This text covers all the preparatory work for modern Galois Theory but stops short of the theory itself. We will also need to refer to irreducibility, multiple roots and symmetric polynomials, all of which were introduced in Section A.1.1.

In Section A.1 we considered polynomials with rational coefficients, i.e. polynomials over the field of rational numbers **Q**. Modern algebra provides a framework for Galois theory in which it can be applied to a wider subject area than rational polynomials although most of our examples in this section will still be rational polynomials.

Through the rest of this chapter we fix a general field K (not necessarily **Q**). However we need to know whether or not K is "like" **Q** in one important sense. Recall

Theorem 1 [ST, pp 3–4]

1. The prime subfield of a field K is the intersection of all subfields of K.

2. Every prime subfield is isomorphic either to the field $\mathbf{Q}$ of rationals or the field $\mathbf{Z_p}$ of integers modulo a prime number p.

3. If the prime subfield of K is isomorphic to $\mathbf{Q}$ we say K has characteristic 0. If the prime subfield of K is isomorphic to $\mathbf{Z_p}$ we say K has characteristic p.

4. If K is a subfield of L then K and L have the same characteristic.

Most of the examples we will see in this section have characteristic 0 and most of the theory is independent of the characteristic. However, there are some theorems that are sensitive to the characteristic and these will be highlighted where they occur.

Galois used the concept of adjoining roots to a given equation and this translates easily into the theory of field extensions. If $f(x) = (x^2 + 1)(x^2 - 5)$ then f is a polynomial over $\mathbf{Q}$ since its coefficients are in $\mathbf{Q}$. We say the factors of f are "irreducible over $\mathbf{Q}$" if there are no elements in $\mathbf{Q}$ that would allow us to reduce f any further. If we "extend" $\mathbf{Q}$ to include $i = \sqrt{-1}$ we denote the field generated (as a subfield of $\mathbf{C}$ say) as $\mathbf{Q}(i)$ which reads "$\mathbf{Q}$ adjoined with i", then

$$f(x) = (x+i)(x-i)(x^2 - 5)$$

as a polynomial over $\mathbf{Q}(i)$. Similarly

$$f(x) = (x+i)(x-i)(x-\sqrt{5})(x+\sqrt{5})$$

as a polynomial over $\mathbf{Q}(i, \sqrt{5})$. We denote this as a *field extension* by $\mathbf{Q}(i, \sqrt{5})$: $\mathbf{Q}$ if we want to make clear what the "ground" field is, being the one we started with. We can classify with

Definition 2 [ST, pp33-34]

1. A simple extension is an extension $L : K$ having the property that $L = K(\alpha)$ for some $\alpha \in L$.

2. Let $K(\alpha) : K$ be a simple extension. If there exists a non-zero polynomial p over K such that $p(\alpha) = 0$ then α is an algebraic element over K and the extension is called a simple algebraic extension. Otherwise α is transcendental over K and $K(\alpha) : K$ is called a simple transcendental extension.

For example, the polynomial $f(x) = a_n x^n + \ldots + a_1 x + a_0$ is an element of a simple transcendental extension of $\mathbf{Q}$, that is, the field of all rational expressions in x. If there is a polynomial p over $\mathbf{Q}$ such that $p(x) = 0$ then $p = 0$.

Field extensions are an important part of Galois Theory since the way we build a field extension, to allow us to solve an equation for its roots, is intimately related to the structure of the Galois group of the equation.

Theorem 3[ST, p35,45,47]:

1. Let $L : K$ be a field extension and suppose that $\alpha \in L$ is algebraic over K. Then the minimum polynomial of α over K is the unique monic polynomial m over K of smallest degree such that $m(\alpha) = 0$.

2. The degree of an extension $L : K$ is denoted $[L : K]$. The degree of a field extension $L : K$ is the dimension of L considered as a vector space over K.

3. If K, L, M are fields and $K \subseteq L \subseteq M$ then $[M : K] = [M : L][L : K]$.

4. Let $K(\alpha) : K$ be a simple extension. If it is transcendental then $[K(\alpha) : K] = \deg m$ where m is the minimum polynomial of α over K.

Examples

1. Consider $\mathbf{C} : \mathbf{R}$. We can construct $\mathbf{C}$ by adjoining $\{1, i\}$ to $\mathbf{R}$ as every element of $\mathbf{C}$ is of the form $x + iy$ where $x, y \in \mathbf{R}$. Then $\{1, i\}$ forms a basis for $\mathbf{C}$ over $\mathbf{R}$ so $[\mathbf{C} : \mathbf{R}] = 2$.

2. The extension $K(x) : K$ where x is transcendental over K has infinite degree since the elements $\{1, x, x^2, \ldots\}$ are linearly independent and form a basis for $K(x)$ over K.

Finally in this section we have the following:
Lemma 4 [ST, pp47–48]

1. A finite extension is one whose degree is finite.

2. An extension $L : K$ is algebraic if every element of L is algebraic over K.

3. $L : K$ is a finite extension if and only if L is algebraic over K and there exist finitely many elements $\alpha_1, \ldots, \alpha_s \in L$ such that $L = K(\alpha_1, \ldots, \alpha_s)$.

A.2.1.2 Splitting Fields. Although formal field theory did not exist in Galois' time he would have had an intuitive notion of fields. When he wrote "... a pth root of unity, α, is included among the quantities that have already been adjoined to the equation."[ED, p108], this would be in modern terminology $\mathbf{Q}(\alpha)$. The concept of a splitting field is quite simple. We construct field extensions over our ground field until we are able to express the given equation in linear factors and, therefore, find the roots of the equation. The splitting field will contain all the roots of f.

For example, in Section A.1 we found that the roots of $u(x) = x^3 - 2$ were z, yz, y^2z where y is a primitive cube root of unity and z the real cube root of 2. We first construct $\mathbf{Q}(y) : \mathbf{Q}$ but find that u does not completely reduce. We then construct $\mathbf{Q}(y, z) : \mathbf{Q}(y)$ so that $u(x) = (x - z)(x - yz)(x - y^2z)$. Our splitting field for u is $\mathbf{Q}(y, z)$.

Theorem 5 [ST,p78–79]: The field Σ is a splitting field for the polynomial f over the field K if $K \subseteq \Sigma$ and

1. f splits into linear factors over Σ.

2. $\Sigma = K(\alpha_1, \ldots, \alpha_n)$ where $\alpha_1, \ldots, \alpha_n$ are the roots of f in Σ.

In the example $u(x) = x^3 - 2$, if we construct the field $\mathbf{Q}(z)$ first then one root of u lies in $\mathbf{Q}(z)$. But $\mathbf{Q}(z) : (Q)$ is not a normal field extension as the other roots of u are not in $\mathbf{Q}(z)$ and u does not split $\mathbf{Q}(z)$.

Definition 7 [ST, p83]:

An irreducible polynomial f over a field K is separable over K if it has no multiple roots in a splitting field. This means that in any splitting field f takes the form
$$f(x) = k(x - \alpha_1) \ldots (x - \alpha_n)$$
where the α_i are all different. An irreducible polynomial over a field K is inseparable over K if it is not separable over K. For example, let $K_0 = Z_p$ where p is prime and let $K = K_0(u)$ where u is transcendental over K_0. If
$$f(x) = x^p - u \in K[x]$$
then let L be a splitting root for f over K. Let τ be a root of f in L so that $\tau^p = u$. We have
$$(x - \tau)^p = x^p + \binom{p}{1} x^{p-1}(-\tau) + \ldots + (-\tau)^p.$$

But p divides $\binom{p}{1}$ for $0 < r < p$ and any multiple of p in K is zero. So $(x - \tau)^p = x^p - \tau^p = x^p - u = f(x)$.

If $\sigma^p - u = 0$ then $(\sigma - u)^p = u$ so $\sigma = \tau$ and all the roots of f in L are equal [ST, p85].

In Section A.1 we showed that any irreducible polynomial with rational coefficients had only simple roots. This result holds for any irreducible polynomial over a field K where K has characteristic 0. However the result is different for fields of characteristic $p > 0$.

Proposition 8 [ST, p86]:

If K is a field of characteristic 0 then every irreducible polynomial f over K is inseparable if and only if

$$f(x) = k_r x^{rp} + \ldots + x^p + k_0 \qquad k_0, \ldots, k_r \in K.$$

If f is defined as above then

$$Df = rpk_r x^{rp-1} + (r-1)pk_{r-1} x^{(r-1)p-1} + \ldots + pk_1 x^{p-1}.$$

Since every multiple of p is zero in a field of characteristic $p > 0$ the result $Df = 0$ does not imply that f is a constant. However, if f has terms that are not raised to a power that is a multiple of p then Df will contain terms which are identically zero only when their coefficient is zero.

The last result in this section is

Lemma 9 [ST, p86]

Let $L : K$ be a separable algebraic extension and let M be an immediate field. Then $M : K$ and $L : M$ are separable.

A.2.2 The Galois Group

A.2.2.1 Automorphisms. If we are able to construct a finite separable normal extension for a given equation we can relate the Galois group of the equation to the field extension through the operation of mappings on the field extension. We can define a mapping from one field to another that satisfies certain conditions. If $\theta : K \to L$ is such a map from the field K to the field L and $x, y \in K$ then

$$\begin{aligned} \theta(x+y) &= \theta(x) + \theta(y) \\ \theta(xy) &= \theta(x)\theta(y). \end{aligned}$$

We shall be concerned with automorphisms which are maps of this type. We are particularly interested in automorphisms which fix our ground field K which are defined as follows

Theorem 10 [ST, p72]

1. If K and L are two fields with $K \subseteq L$ and $\theta : L \to L$ is an automorphism such that $\theta(k) = k$ for all $k \in K$ then θ is a K-automorphism of L.

2. The set of all K-automorphisms of L forms a group under composition of maps.

3. The Galois group $\Gamma(L : K)$ of the extension $L : K$ is the group of all K-automorphisms of L under composition of maps.

Thus we have a connection between a field extension and its Galois group.

Proof(of(2))

If α and β are K-automorphisms of L then $\alpha\beta$ is also an automorphism. For $k \in K$
$$\alpha\beta(k) = \alpha(k) \Rightarrow \alpha\beta \text{ is a } K\text{-automorphism.}$$
If e is the identity then e is a K-automorphism since $e(k) = e$
$$\alpha^{-1} \text{ is an automorphism of } L \text{ and } k = \alpha^{-1}\alpha(k) = \alpha^{-1}(k) \Rightarrow \alpha^{-1}$$
is an automorphism of L.

□

Let us return again to the example $u(x) = x^3 - 2$. Suppose the field extension is $\mathbf{Q}(y, z) : \mathbf{Q}$ then we have the following possible $\mathbf{Q}$-automorphisms of $\mathbf{Q}(y, z)$.

$$\theta : \quad \theta(y) = y \quad \theta(z) = yz, \quad \theta^3 = e$$
$$\phi : \quad \phi(y) = y^2 \quad \phi(z) = z, \quad \phi^2 = e$$

We can construct the following table which gives all possible $\mathbf{Q}$-automorphisms of $\mathbf{Q}(y, z)$

(Q)-automorphism	Effect on y	Effect on z
c	y	z
θ	y	yz
θ^2	y	y^2z
ϕ	y^2	z
$\phi\theta^2 = \theta\phi$	y^2	yz
$\phi\theta = \theta^2\phi$	y^2	y^2z

So $\Gamma(\mathbf{Q}(x,y) : \mathbf{Q}) = \{e, \theta, \theta^2, \phi, \theta\phi, \theta^2\phi\}$. This group is isomorphic to S_3 which is the Galois group of u that we found in Section A.1.

A.2.2.2 Subfields and Subgroups. The elements of the Galois group relate to the splitting field by defining subfields. Each subfield has its own automorphism in the larger field and these also form a group. The automorphisms of a splitting field are generally the permutations of roots such as we considered in Section A.1.2. Subfields may be considered to be intermediate stages in the adjunction of roots although the process is somewhat different in this context.

A. Project Example 1: Topics in Galois Theory

Definition 11 [ST, pp73–75]

If $L : K$ is a field extension let M be an "intermediate" field such that $K \subseteq M \subseteq L$. To each intermediate field M we associate a group $M*$ which contains all the M-automorphisms of the field L. $\Gamma(L : M)$ is a subset of the Galois group $\Gamma(L : K)$. Now K^* is the group of all K-automorphisms of L and is therefore the whole Galois group, whilst L^* contains only the identity automorphism. If $M \subseteq N$ then $M^* \supseteq N^*$ since any automorphism fixing the elements of N will fix the elements of M.

Working from the other direction, given a subgroup of $\Gamma(L : K)$ we can associate with it a set of fixed elements of the field L. If H is a subgroup of $\Gamma(L : K)$ then $\hat{H}$ denotes the set of all elements in L fixed by the automorphisms in H.

The relationship between the subgroups and subfields is given by

Lemma 12 [ST, p74]

If H is a subgroup of $\Gamma(L : K)$ then $\hat{H}$ is a subfield of L containing K. We call $\hat{H}$ the fixed field of H. If $H \subseteq G$ then $\hat{H} \supseteq \hat{G}$ since any elements fixed by G will be fixed by H. If M is an intermediate field and H is a subgroup of the Galois group then $M \subseteq \hat{M^*}$ and $H \subseteq \hat{\hat{H}}^*$.

The example $u(x) = x^3 - 2$ has the following fixed fields and related subgroups:

Subgroup of $\Gamma(\mathbf{Q}(y,z) : \mathbf{Q})$	Fixed field
e	$\mathbf{Q}(y,z)$
$\{e, \theta^2\phi\}$	$\mathbf{Q}(yz)$
$\{e, \theta\phi\}$	$\mathbf{Q}(yz(y+z))$
$\{e, \phi\}$	$\mathbf{Q}(z)$
$\{e, \theta, \theta^2\}$	$\mathbf{Q}(y)$
$\{e, \theta, \theta^2, \phi, \theta\phi, \theta^2\phi\}$	$\mathbf{Q}$

A more complicated example is the field extension $\mathbf{Q}(\sqrt{2}, \sqrt{3}, \sqrt{5}) : \mathbf{Q}$. Let this be $L : \mathbf{Q}$. Let

$$r_1 = \sqrt{2},\ r_2 = -\sqrt{2},\ r_3 = \sqrt{3},\ r_4 = -\sqrt{3},\ r_5 = \sqrt{5},\ r_6 = -\sqrt{5}$$

then the $\mathbf{Q}$-automorphisms can be represented as permutations on the r_i. If θ is a permutation then $\theta(r_i) = r_{\theta(i)}$. Our Galois group is then

$$\Gamma(L : \mathbf{Q}) = \{e, (12), (34), (56), (12)(56), (34)(56), (12)(34)(56)\}.$$

Some proper subgroups of $\Gamma(L : \mathbf{Q})$ are

$$B = \{e, (12)(34), (12)(56), (12)(56)\},\ C = \{e, (12), (34), (12)(34)\},$$
$$D = \{e, (12), (56), (12)(56)\},\ E = \{e, (34), (56), (34)(56)\},$$
$$F = \{e, (12)(34)\},\ G = \{e, (12)(56)\},\ H = \{e, (34)(56)\},$$
$$R = \{e, (12)\},\ S = \{e, (34)\},\ T = \{e, (56)\},\ U = \{e, (12)(34)(56)\}.$$

The fixed fields associated with these subgroups are

$$\begin{aligned}\hat{B} &= \mathbf{Q}, \quad \hat{C} = \mathbf{Q}(\sqrt{5}), \quad \hat{D} = \mathbf{Q}(\sqrt{3}), \quad \hat{E} = \mathbf{Q}(\sqrt{2}), \\ \hat{F} &= \mathbf{Q}(\sqrt{5}, \sqrt{6}), \quad \hat{G} = \mathbf{Q}(\sqrt{3}, \sqrt{10}), \quad \hat{H} = \mathbf{Q}(\sqrt{2}, \sqrt{15}), \\ \hat{R} &= \mathbf{Q}(\sqrt{3}, \sqrt{5}), \quad \hat{S} = \mathbf{Q}(\sqrt{2}, \sqrt{5}), \quad \hat{T} = \mathbf{Q}(\sqrt{2}, \sqrt{3}), \\ \hat{U} &= \mathbf{Q}(\sqrt{6}, \sqrt{10}, \sqrt{15}).\end{aligned}$$

We also have $\hat{e} = K$ and $\Gamma(K : \mathbf{Q})\hat{\,} = \mathbf{Q}$.

In summary we have

Definition 13 [ST, p74]

If $\mathcal{F}$ denotes the set of intermediate fields, and $\mathcal{G}$ the set of subgroups of the Galois group then we have two maps

$$\begin{aligned} * : \mathcal{F} &\to \mathcal{G} \\ \hat{\,} : \mathcal{G} &\to \mathcal{F} \end{aligned}$$

which reverses inclusions and satisfies Lemma 12.

A.2.2.3 Group Order and Degree of Field Extension. There is also a relationship between the degree of a field extension and the order of the group of automorphisms associated with the extension.

Theorem 14 [ST, p91]

1. Let G be a finite subgroup of the group of automorphisms of a field K, and let K_0 be the fixed field of G. Then $[K : K_0] = |G|$.

2. If G is the Galois group of the finite extension $L : K$ and H is a finite subgroup of G then $[\hat{H}:K] = [L : K]/|H|$.

Consider the example of $\mathbf{Q}(\sqrt{2}, \sqrt{3}, \sqrt{5}) : \mathbf{Q} = L : \mathbf{Q}$ detailed above.

$$\begin{aligned}[L : \mathbf{Q}] &= [L : \mathbf{Q}(\sqrt{2}, \sqrt{3})][\mathbf{Q}(\sqrt{2}, \sqrt{3}) : \mathbf{Q}(\sqrt{2})][\mathbf{Q}(\sqrt{2}) : \mathbf{Q}] \\ &= 2.2.2 = 8.\end{aligned}$$

Let $Z = \gamma(L : \mathbf{Q})$ then $|Z| = 8$. We have $Z \supset S = e, (34)$ and $|S| = 2$. Now the fixed field of S is $\hat{S} = \mathbf{Q}(\sqrt{2}, \sqrt{5})$ and $[\hat{S}:\mathbf{Q}] = [\mathbf{Q}(\sqrt{2}, \sqrt{5}) : \mathbf{Q}(\sqrt{2})][\mathbf{Q}(\sqrt{2}) : \mathbf{Q}] = 2.2 = 4$. So $4 = [\hat{S} : \mathbf{Q}] = [L : \mathbf{Q}]/|S| = 8/2 = 4$.

Another example is supplied by the field extension for $u(x) = x^3 - 2$.

Let $G = \Gamma(\mathbf{Q}(y, z) : \mathbf{Q})$ and $H = \{e, \theta, \theta^2\}$ so $|H| = 3$.

Also $[\mathbf{Q}(y, z) : \mathbf{Q}] = [\mathbf{Q}(y, z) : \mathbf{Q}(y)][\mathbf{Q}(y, z) : \mathbf{Q}] = 3.2 = 6$.

The fixed field of H is $\hat{H} = \mathbf{Q}(y)$, $[\hat{H} : \mathbf{Q}] = 2$. So

$$2 = [\hat{H} : \mathbf{Q}] = [\mathbf{Q}(y, z) : \mathbf{Q}]/|H| = 6/3 = 2.$$

In the above two examples the degree of the field extension that gives the splitting field is equal to the order of the Galois group of the extension. This is formally stated with

Theorem 15 [ST, p100]

If $L : K$ is a finite separable normal extension of degree n then there are precisely n distinct K-automorphism of L, so that $|\Gamma(L : K)| = n$.

Suppose that a separable extension $L : K$ is not normal. It can be shown [ED, pp98–102] that a 'normal closure' that enlarges the extension $L : K$ can be constructed, and that Theorem 15 will still hold for $L : K$. A normal closure is simply an extension N of L such that $N : K$ is normal, and N is the smallest such extension of L which is normal over K.

A.2.3 Applying Galois Theory

A.2.3.1 The Galois Correspondence. We can now give a formal statement of the relationship between a field extension and its Galois group. This is known as the Galois correspondence and is the fundamental theorem of Galois theory in its modern algebra setting.

Theorem 16 [ST pp104–105]

If $L : K$ is a finite separable normal field extension of degree n, with Galois group G and if $\mathcal{F}, \mathcal{G}, *, \hat{\ }$ are defined as in Section A.2.2 then

1. The Galois group has order n.

2. The maps $*$ and $\hat{\ }$ are mutual inverses and set up an order reversing $1-1$ correspondence between $\mathcal{F}$ and $\mathcal{G}$.

3. If M is an intermediate field then $[L : M] = |M^*|$ and $[M : K] = |G|/|M^*|$.

4. An intermediate field M is a normal extension of K if and only if M^* is a normal subgroup of G.

5. If an intermediate field M is a normal extension of K then the Galois group of $M : K$ is isomorphic to the quotient group G/M^*.

We can show that the Galois correspondence of $\mathbf{Q}(\sqrt{2}, \sqrt{3}, \sqrt{5}) : \mathbf{Q} = L : \mathbf{Q}$ as follows (refer to Section A.2.2 for the details):

1. $[L : \mathbf{Q}] = 8 = |Z|$.

2. $L = \hat{e}$, $\mathbf{Q}(\sqrt{2}, \sqrt{3}) = \hat{T}$, $\mathbf{Q}(\sqrt{2}) = \hat{E}$, $\mathbf{Q} = \hat{Z}$, $e = L^*$, $T = \mathbf{Q}(\sqrt{2}, \sqrt{3})^*$, $E = \mathbf{Q}(\sqrt{2})$, $Z = \mathbf{Q}^*$.

3. $[L : \mathbf{Q}(\sqrt{2}, \sqrt{3})] = 2 = |T|$, $[\mathbf{Q}(\sqrt{2}, \sqrt{3}) : \mathbf{Q}] = 4 = |Z|/|T|$, $[L : \mathbf{Q}(\sqrt{2})] = 4 = |E|$, $[\mathbf{Q}(\sqrt{2}) : \mathbf{Q}] = 2 = |Z|/|E|$.

4. Z is abelian so every subgroup of Z is normal.
   Therefore $\mathbf{Q}(\sqrt{2}) : \mathbf{Q}$, $\mathbf{Q}(\sqrt{2}, \sqrt{3}) : \mathbf{Q}(\sqrt{2})$ and $L : \mathbf{Q}(\sqrt{2}, \sqrt{3})$ are all normal extensions.

5. Let $V = \{(12), (12)(34), (12)(56), (12)(34)(56)\}$. Then $Z/E = \{E, V\}$ which is isomorphic to $\Gamma(\mathbf{Q}(\sqrt{2}) : \mathbf{Q}) = R = \{e, (12)\}$. Let

$$W = \{(12), (12)(56)\}, \quad X = \{(34), (34)(56)\}, \quad Y = \{(12)(34), (12)(34)(56)\}$$

then $Z/T = \{T, W, X, Y\}$ is isomorphic to $\Gamma(\mathbf{Q}(\sqrt{2}, \sqrt{3}) : \mathbf{Q}) = C = \{e, (12), (34), (12)(34)\}$.

The field extension for $u(x) = x^3 - 2$ also satisfies the Galois correspondence, with $\mathbf{Q}(y)$ being an intermediate field. $\mathbf{Q}(y)^* = \{e, \theta, \theta^2\} \cong A_3$ and A_3 is a normal subgroup of S_3.

A.2.3.2 Radical Extensions. Although field extensions are the 'adjunction of quantities' that Galois proposed, the modern expression of Galois Theory relies on *starting* with a splitting field and analysing the behaviour of the Galois group as elements of the splitting field are fixed. In practice, as we are unaware of the elements in the splitting field, we rely entirely on studying the nature of this field and its associated Galois group. That is, ensuring that our splitting field satisfies the conditions in Theorem 16 and analysing the solubility of the group. However, we have not yet imposed any conditions on the em type of elements by which we extend the ground field. If we wish to show that a polynomial can be solved, then we must have a "radical extension".

Definition 17 [ST, p128]

An extension $L : K$ is radical if $L = K(\alpha_1, \ldots, \alpha_n)$ where for each $i = 1, \ldots, m$ there exists an integer $n(i)$ such that

$$\alpha_i^{n(i)} \in K(\alpha_1, \ldots, \alpha_{i-1}) \qquad i \geq 2.$$

The elements α_i are said to form a radical sequence for $L : K$.

For example if $L : \mathbf{Q}(\sqrt{2}, \sqrt{3})$ then $(\sqrt{3})^2 = 3 \in \mathbf{Q}(\sqrt{2})$.

So a radical extension is nothing more than a series of adjunctions of pth roots as detailed by Galois. It is at this point we concentrate on fields of characteristic 0 since the proofs of some of the following items contain this assumption.

Theorem 18[ST, pp129–130]

1. If $L : K$ is a radical extension and M is a normal closure of $L : K$ then $M : K$ is radical.

2. If K has characteristic 0 and $L : K$ is normal and radical then $\Gamma(L : K)$ is soluble.

3. If K is a field of characteristic 0 and $K \subseteq L \subseteq M$ where $L : K$ is a soluble group.

Consider $\mathbf{Q}(\sqrt{2}, \sqrt{3}, \sqrt{5}) : \mathbf{Q}$. This is, by definition, a radical extension. Therefore we should expect its Galois group, Z, to be soluble. This is confirmed in Section A.1.4

$\mathbf{Q}(\sqrt{2}, \sqrt{3}) : \mathbf{Q}$ is less obvious. We know that y and z are expressible in radicals so the splitting field must be a radical extension. In general, however, we may not know whether an extension is radical. In such a case we can turn to a theorem which is almost the converse of Theorem 18.

Theorem 19[ST, p146]

Let K be a field of characteristic 0 and let $L : K$ be a finite normal extension with soluble Galois group G. Then there exists an extension R of L such that $R : K$ is radical.

So, since we have already shown that S_3 is soluble, we can state the extension $\mathbf{Q}(y, z) : \mathbf{Q}$ is radical and u is soluble by radicals.

For fields of characteristic $p > 0$ we must impose an additional condition to ensure that a group implies an equation soluble by radicals. In addition to the adjoining of radical elements as defined in Definition 17 we must also adjoin elements α such that $\alpha^p - \alpha$ is in the given field where p is the characteristic. The adjunction of such elements extends the proof of Theorem 19 to allow inclusion of fields of characteristic $p > 0$. Otherwise, we may have some polynomials with soluble Galois group that are not soluble by radicals.

A.2.3.3 The General Polynomial of degree n. We have so far considered only finite algebraic extensions. To tackle the general polynomial equation

$$g(x) = x^n - \sigma_1 x^{n-1} + \ldots + (-1)^n \sigma_n$$

(where the coefficients are written in terms of the elementary symmetry polynomials σ_k) we will need to use a transcendental extension.

Definition 20[ST, p139]

1. An extension $L : K$ is finitely generated if $L = K(\alpha_1, \ldots, \alpha_n)$ where n is finite. This is true whether the α_i are algebraic or transcendental over K.

2. If $t_1, \ldots, t_n$ are transcendental elements over a field K, all lying inside some extension L of K then they are independent if there is no non-trivial polynomial p over K (in n indeterminates) such that $p(t_1, \ldots, t_n) = 0$ in L.

3. Let $t_1, \ldots, t_n$ be independent transcendental elements over K. The symmetric group S_n can be made to act as a group of K-automorphisms of $K(t_1, \ldots, t_n)$ by defining $\theta(t_i) = t_{\theta(i)}$ for all $\theta \in S_n$. Distinct elements give rise to distinct K-automorphisms.

As we would expect, the fixed field of S_n consists of the quotients of the symmetry polynomials in the t_i. In particular it contains the elementary symmetric polynomials σ_k, so if F is the fixed field of S_n we have $F = K(\sigma_1, \ldots, \sigma_n)$.

Lemma 21[ST, p143]

With the above notation, $\sigma_1, \ldots, \sigma_n$ are independent transcendental elements over K.

Now, if we consider the t_i to be the n roots of the general polynomial of degree n over $K(\sigma_1, \ldots, \sigma_n)$ where

$$g(x) = x^n - \sigma_1 x^{n-1} + \ldots + (-1)^n \sigma_n.$$

Thus we have

Theorem 22[ST, p143]

For any field K let g be the general polynomial of degree n over K and let Σ be a splitting field for g over $K(\sigma_1, \ldots, \sigma_n)$. Then the roots $t_1, \ldots, t_n$ of g in Σ are independent transcendental elements over K, and the Galois group of $\Sigma : K(\sigma_1, \ldots, \sigma_n)$ is the full symmetric group S_n.

Corollary 23

The general polynomial of degree n has Galois group S_n which is not soluble for $n \geq 5$.

A.3 Soluble Groups

A.3.1 Normal Subgroup Series

In Section A.1 we showed that a soluble group has a series of normal subgroups of index p (for p prime) and this series terminates with the identity element.

A more general definition is available which does not confine solubility to the action of adjoining pth roots.

Definition 1 [ST, pp 114–115]
A group B is soluble if it has a finite series of subgroups

$$B = B_0 \supseteq B_1 \supseteq \ldots \supseteq B_n = e$$

called a composition series, such that

1. $B_{i+1} \triangleleft B_i$ for $i = 0, \ldots, n-1$;
2. B_i/B_{i+1} is abelian for $i = 0, \ldots, n-1$.

Subgroup normality is not transitive so $B_i \triangleleft B_{i-1} \triangleleft B_{i-2} \neq B_i \triangleleft B_{i-2}$.

Note that if we compare this to Galois' definition of a soluble group then $|B_i/B_{i+1}| = p$ for a prime number p. Therefore B_i/B_{i+1} is abelian since all groups of prime order are cyclic and hence abelian.

We know that from the examples in Section A.1.2 that the decomposition of a group into subgroups is not necessarily unique. If we have more than one way of composing a series of normal subgroups, we need to be sure that each different series satisfies the definition of a soluble group. The Jordan–Holder Theorem gives

Theorem 2[AL, p263]
Let B be a finite group with composition series

$$\begin{aligned} B &= B_0 \supseteq B_1 \supseteq \ldots \supseteq B_n = e \\ \text{and} \quad B &= C_0 \supseteq C_1 \supseteq \ldots \supseteq C_m = e. \end{aligned}$$

Then $m = n$ and the r quotient groups B_i/B_{i+1} can be put in 1–1 correspondence with the r quotient groups C_i/C_{i+1} in such a manner that the corresponding groups are isomorphic. We will not necessarily have $B_i/B_{i+1} \cong C_i/C_{i+1}$ but all the quotient groups can be paired off. Provided we can construct *one* series that satisfies Definition 1 then every series will also satisfy the definition.

A.3.2 Normal Subgroups

As we have seen in Section A.3.1 the solubility of a group depends on the normality of the subgroups. Abelian groups are useful since every subgroup of an abelian group is normal, and the quotient of an abelian group is always normal. This means that every abelian group is soluble and therefore every cyclic group is soluble since they are all abelian.

In Section A.1 we mentioned that one of Galois' examples of a soluble equation was the cyclotomic equation

$$\frac{x^p - 1}{x - 1} = x^{p-1} + \ldots + x + 1$$

where p is prime. It is clear that one of the roots of $x^p - 1$ is 1 so we are interested only in the roots of $x^{p-1} + \ldots + x + 1$. This is an irreducible equation with simple roots. Note that if a is one of these roots then all the roots can be expressed as a^i for $i = 1, \ldots, p-1$. So the splitting field for this equation is $\mathbf{Q}(a)$ and the $\mathbf{Q}$-automorphisms are

$$\begin{aligned}
\phi_1 : \quad & a \rightarrow a \\
\phi_2 : \quad & a \rightarrow a^2 \\
\phi_3 : \quad & a \rightarrow a^3 \\
& \vdots \quad\quad \vdots \\
\phi_{p-1} : \quad & a \rightarrow a^{p-1}.
\end{aligned}$$

The Galois group is abelian since for any two $\mathbf{Q}$-automorphisms we have

$$\phi_i(a)\phi_j(a) = a^i a^j = a^{i+j} = a^j a^i = \phi_j(a)\phi_i(a)$$

so the Galois group is soluble, $\mathbf{Q}(a)$ is a radical extension and all pth roots of unity are expressible in radicals when p is prime. Since we have already shown that any nth root, where n is not prime, can be constructed by using pth roots when p is a prime number, this shows that the cyclotomic equation $x^n - 1$ is soluble by radicals for any n. The following theorem is useful when we have partial knowledge of the structure of a group.

Theorem 3 [ST, p116]

Let G be a group, H a subgroup of G, and N a normal subgroup of G.

1. If G is soluble then H is soluble;

2. If G is soluble then G/N is soluble;

3. If N and G/N is soluble then G is soluble.

Every subgroup H of a group G, where H has index 2 in G is a normal subgroup. A particular family of such subgroups is the alternating group A_n which has index 2 in the symmetric group S_n. So A_n is always a normal subgroup S_n. However, for $n \geq 5$ this information, as we might expect, is not much help. This leads us on to our next section.

A.3.3 Simple Groups

Theorem 4 [ST, pp117–118]

1. A group G is simple if its only normal subgroups are e and G.

2. A soluble group is simple if and only if it is cyclic of prime order.

3. If $n \geq 5$ then the alternating group A_n of degree n is simple.

The importance of this last result is highlighted by

Corollary 5 [ST, p119]
The symmetric group S_n of degree n is not soluble if $n \geq 5$.

Proof
If S_n were soluble then A_n would be soluble. But if $n \geq 5$ A_n is simple and of order $n!/2$, which is not prime for $n \geq 5$, and therefore not soluble. This shows that for $n \geq 5$ S_n is not soluble.

□

A.3.4 p-Groups

These are another example of soluble groups. We have

Lemma 6[ST, p121]:

1. Let p be a prime. A finite group B is a p-group if its order is a power of p.

2. If B is a finite p-group of order p^n then B has a series of normal subgroups
$$B = B_0 \supseteq B_1 \supseteq \ldots \supseteq B_n = e$$
such that $|B_i| = p^{n-i}$ for all $i = 0, \ldots, n$.

3. Every finite p-group is soluble.

Proof(of (3))
Every quotient group has order
$$|B_i|/|B_{i+1}| = p^{n-i}/p^{n-(i+1)} = p$$
which is prime. Therefore every quotient group is cyclic and abelian. As every subgroup is normal, this implies B is soluble.

□

The theory of p-groups was extended by Sylow who gave the following theorem

Theorem 7 [ST, p123]

Let B be a finite group of order $p^a r$ where p is prime and does not divide r. Then

1. B possess at least one subgroup of order p^a;

2. all such subgroups are conjugate in B;

3. any p-subgroup of B is contained in one of order p^a;

4. the number of subgroups of B of order p^a leaves remainder 1 on division by p.

If the Sylow p-subgroup P of B is a normal subgroup of B, and B/P is abelian, then G is soluble, since P is soluble.

S_5 is a group of order $120 = 2^3.3.5$. So S_5 should have Sylow p-subgroups for $p = 2$, $p = 3$ and $p = 5$. Let P_2 be a Sylow 2-subgroup of S_5 then $|P_2| = 8$. Using similar notation we also have $|P_3| = 3$ and $|P_5| = 5$.

Allenby gives a more detailed version of part 4 in Theorem 7:

4 The number of conjugates of a Sylow p-subgroup of B divides $|B|/p^a$ and is congruent to $1 (\mod p)$

Consider S_3 which has order $6 = 2.3$. S_3 has one Sylow 3-subgroup which is A_3.
$$S_3/3 = 2 \text{ and } 1|2 \qquad 1 = 1(\mod 3).$$
S_3 has a Sylow 2-subgroup which is $X = \{e, (12)\}$ where $X \cong \{e, (13)\} \cong \{e, (23)\}$
$$S_3/2 = 3 \text{ and } 3|3 \qquad 3 = 1(\mod 2).$$
Both A_3 and X are normal in S_3.

The example from Section A.2 of $Z = \Gamma(\mathbf{Q}(\sqrt{2}, \sqrt{3}, \sqrt{5}) : \mathbf{Q})$ is an example of a p-group since $|Z| = 8 = 2^3$. So, as we expected, Z is a soluble group.

A.4 Geometrical Constructions

In this section we study an interesting application of field theory to classical geometry. Since ancient Greek times mathematicians have been interested in the construction of geometrical figures using only a straight edge and compass. With this restriction we are allowed two operations:

1. drawing a line through two given points;

2. drawing a circle with centre at one point and passing through a second given point.

Many figures have proved to be constructible under these restrictions but not all. Three problems were of particular interest to the ancient Greeks and subsequent generations of mathematicians over the following centuries. These problems are [DU, p264]:

1. The Duplication of the Cube–given a cube construct another cube with twice the volume of the given cube.

2. The Trisection of an Arbitrary Angle–some angles can be trisected but a method is required for the trisection of *any* angle.

3. The Quadrature of the Circle–given a circle construct the square that has the same area as the given circle.

Not until the 19th century was it proved that these three problems have no solution. The "impossibility proofs" depend on algebraic techniques rather that geometry. We shall use field theory, particularly that of field extensions, to prove the impossibility of these three constructions.

A.4.1 Constructible Points

We need to find a way of expressing geometrical construction in algebraic terms. We can assume that we are given at least two points. Let the set of given points be $P_0 \subset \mathbf{R}^2$ and let $p_1, p_2 \in P_0$. Then we can use our two operations to construct:

1. a line through p_1 and p_2;

2. a circle with centre p_1 and radius $p_1 p_2$;

3. a circle with centre p_2 and radius $p_1 p_2$.

We add points to P_0 by adjoining the intersection points of such lines and circles. Let r_1 be one of these intersection points. Then r_1 is *constructible* in one step from P_0. We define a point r in $\mathbf{R}^2$ to be constructible from P_0 if there is a finite sequence $r_1, r_2, \ldots, r_n = r$ of points in $\mathbf{R}^2$ such that for each $i = 1, 2, \ldots, n$ the point r_i is constructible in one step from the set $P_0 \cup \{r_1, \ldots, r_{i-1}\}$[ST, p52].

Let K_0 be the subfield of $\mathbf{R}$ generated by the x and y coordinates of the points in P_0. Let r_i be an intersection point with coordinates (x_i, y_i). Then

K_i is the subfield of $\mathbf{R}$ generated by adjoining x_i and y_i with K_{i-1} i.e. $K_i = K_{i-1}(x_i, y_i)$. So
$$K_0 \subseteq K_1 \subseteq \ldots \subseteq K_n \subseteq \mathbf{R}.$$
With this notation we have

Lemma 1 [ST, p53]

x_i and y_i are roots in K_i of quadratic polynomials over K_{i-1}.

Outline proof

The point r_i is an intersection point of either two lines, two circles, or a line and a circle all constructed from points of P_{i-1}. If r_i is the intersection point of

1. *Two lines* then $x_i, y_i \in K_{i-1}$ which implies $x_i, y_i \in K_i$. Also x_i and y_i are certainly roots of quadratics polynomials over K_{i-1};

2. *Two circles* then we obtain quadratic expressions for x and y with coefficients in K_{i-1};

3. *A circle and a line* then again we have quadratic expressions for x and y with the same result as in (2) above.

The following theorem is the method by which we will prove the impossibility of the three constructions mentioned at the start of the section.

Theorem 2 [ST, pp54–55]

If $r = (x, y)$ is constructible from a subset P_0 of $\mathbf{R}^2$ and if K_0 is the subfield of $\mathbf{R}$ generated by the coordinates of the points of P_0, then the degrees $[K_0(x) : K_0]$ and $[K_0(y) : K_0]$ are powers of 2.

Proof

If r_i is the intersection of two lines then $x_i, y_i \in K_{i-1}$ and $[K_{i-1}(x_i) : K_{i-1}] = [K_{i-1}(y_i) : K_{i-1}] = 1$

If r_i is the intersection of two circles or of a circle and a line then x_i and y_i are the roots of quadratic polynomials over K_{i-1}. If the quadratic polynomials are reducible then the result is the same as for the intersection of two lines. If the quadratic polynomials for x_i and y_i are irreducible then
$$[K_{i-1}(x_i) : K_{i-1}] = [K_{i-1}(y_i) : K_{i-1}] = 2.$$
So we can write
$$[K_{i-1}(x_i, y_i) : K_{i-1}] = [K_{i-1}(y_i) : K_{i-1}(x_i)][K_{i-1}(x_i) : K_{i-1}] = 1, 2 \text{ and } 4$$

If $[K_i : K_{i-1}]$ is a power of 2 then by induction $[K_n : K_0]$ is a power of 2. But
$$[K_n : K_0(x)] = [K_0(x) : K_0] = [K_n : K_0]$$
so $[K_0(x) : K_0]$ is also a power of 2. Similarly, $[K_0(y) : K_0]$ is a power of 2 [ST, p55].

$\square$

A.4.2 Impossibility Proofs

We can assume that our given points P_0 include a pair of perpendicular axes and the unit circle so that $\{(0,0),(1,0)\} \in P_0$. The following three proofs show that it is not possible to construct these figures using only straight edge and compass.

1. *Duplication of the Cube* Without loss of generality, we can assume that the given cube is the unit cube and that one side is the segment of the x-axis $(0,0)$ to $(1,0)$. Hence we wish to construct the cube with volume 2 units and with length of side α such that $\alpha^3 = 2$. If we could construct the point $(\alpha, 0)$ then $[\mathbf{Q}(\alpha) : \mathbf{Q}]$ would be a power of 2. But α has a minimum polynomial $x^3 - 2$ over $\mathbf{Q}$ which is irreducible over $\mathbf{Q}$. So $[\mathbf{Q}(\alpha) : \mathbf{Q}] = 3$ and therefore $(\alpha, 0)$ is not constructible [ST, pp55–56].

2. *Trisection of an Arbitrary Angle* We will show that the angle $\pi/3$ cannot be trisected. We wish to construct the angle $\pi/9$ which is equivalent to constructing the point $(\alpha, 0)$ where $\alpha = \cos(\pi/9)$. Recall that for any angle A, $\cos A = 4\cos^3(A/3) - 3\cos(A/3)$, so if A is $\pi/3$ this gives
$$4\cos^3(\pi/9) - 3\cos(\pi/9) = 1/2$$
$$8\alpha^3 - 6\alpha - 1 = 0.$$

Now $8x^3 - 6x - 1$ is irreducible over $\mathbf{Q}$ therefore $\mathbf{Q}(\alpha) : \mathbf{Q} = 3$. This contradiction shows that $(\alpha, 0)$ cannot be constructed ([DU, p265] and [ST, p56]).

3. *Quadrature of the Circle* Without loss of generality we can assume that the given circle is the unit circle. So we wish to construct the square with area π. This is equivalent to constructing the point $(\sqrt{\pi}, 0)$. If we can construct $(\sqrt{\pi}, 0)$ then we can certainly construct $(\pi, 0)$ (constructing the square of a number is demonstrated in Section A.1.3) so that $[\mathbf{Q}(\pi) : \mathbf{Q}] = 2^m$ for some $m \geq 0$. But π is not algebraic over $\mathbf{Q}$ (by Lindermann's Theorem) and therefore $[\mathbf{Q}(\pi) : \mathbf{Q}] \neq 2^m$ which shows that we cannot construct $(\pi, 0)$ and therefore by implication $(\sqrt{\pi}, 0)$ [ST, p57].

A.4.3 Performing Algebraic Operations by Construction

We can now give demonstrations of constructions where we are carrying out an algebraic operation using the method of geometrical construction.

1 *Constructing the square of a number* Given the point a we construct a^2 by drawing the line from $(0, a)$ to $(1, 0)$ then constructing the line parallel to this line and passing through $(a, 0)$. By similar triangles the point $(0, b)$ at which the line intersects the vertical axis is a^2 since

$$\frac{a}{1} = \frac{b}{a}.$$

2 *Constructing the square root of a number*: Given the point b we construct $\sqrt{b}$ by drawing the circle which has diameter $(-1, 0)$ to $(b, 0)$. This circle will intersect the vertical axis at $(0, a)$. By the intersecting chords theorem $a^2 = 1 \times b$ so that $a = \sqrt{b}$.

A.4.4 Regular n-gons

In 1796, at the age of 18, Gauss found a construction by straight edge and compass for the regular 17-gon. This was the first new such polygon for 2,000 years. Prior to this it was known how to construct regular n-gons for only the following values of n [AL, p174]:

$$2^r, \qquad 2^r.3, \qquad 2^r.5, \qquad 2^r.15.$$

In 1801 Gauss gave the following theorem:

Theorem [ST, p169]

The regular n-gon is constructible by straight edge and compass if and only if $n = 2^r p_1 \ldots p_s$ where r and s are integers greater than 0 and $p_1 \ldots p_s$ are odd primes of the form $p_i = 2^{2^n} + 1$ for positive integers r_i (proof: [ST, pp169–170]).

The p_i are known as Fermat numbers since they were discovered by Fermat in the 17th century. If $F_n = 2^{2^n} + 1$ then $F_0 = 3$, $F_1 = 5$, $F_2 = 17$, $F_3 = 257$, $F_4 = 65,537$. These are all primes and Fermat conjectured that F_n is prime for all n. This was disproved in 1732 by Euler when he showed that F_5 is not a prime. This leads us to

Proposition [ST, p170]

The only primes $p < 10^{40,000}$ for which the regular p-gon is constructible are 2, 3, 5, 17, 257, 65537.

A. Project Example 1: Topics in Galois Theory

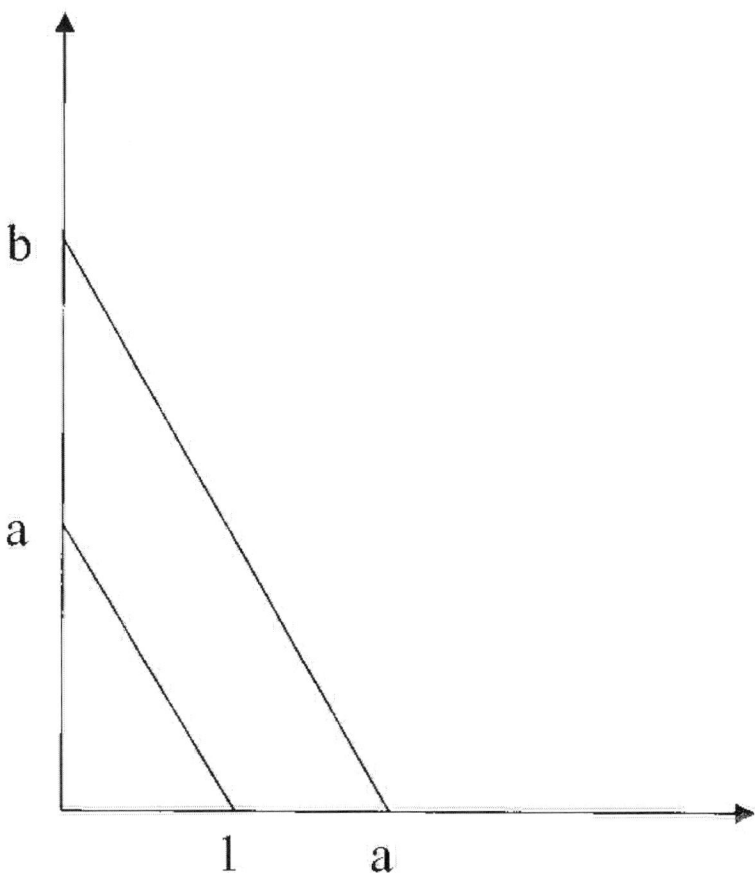

Figure A.1   Constructing the square of a number

If we wish to construct a p-gon where p is prime we can make use of the cyclotomic equation
$$\frac{x^p - 1}{x - 1} = x^{p-1} + \ldots + x + 1.$$
The roots of the right hand side of the equation are a_k where $a_k = \cos(2k\pi/p) + i\sin(2k\pi/p)$.

As an example we will find a construction for the regular pentagon. If we assume that the pentagon has centre $(0,0)$ and its first vertex is at $(1,0)$ then we wish to construct the angle $\theta = 2\pi/5$ which is equivalent to constructing

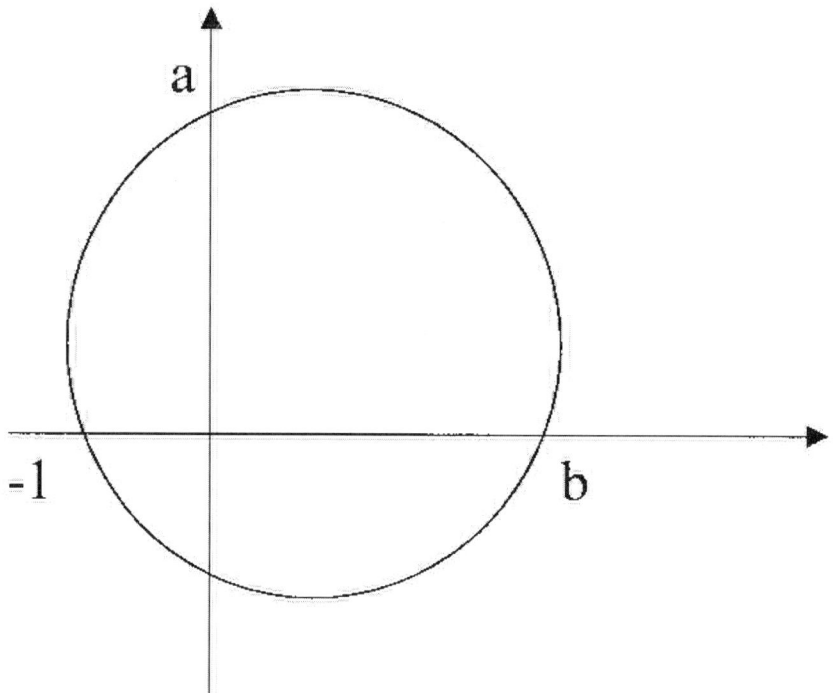

Figure A.2   Constructing the square root of a number

the point $(\alpha, 0)$ where $\alpha = \cos\theta$.

The primitive roots of $x^5 - 1$ are a_1, a_2, a_3, a_4 where $a_1 + a_2 + a_3 + a_4 = -1$ (an elementary symmetric polynomial). We also need

$$\begin{aligned}
a_{5-k} + a_k &= \cos(5-k)\theta + \cos k\theta + i\sin(5-k)\theta + i\sin k\theta \\
&= 2\cos(2\pi)\cos\left[\frac{(5-2k)\pi}{5}\right] + 0 \\
&= 2\cos(2k\pi/5) \\
&= 2\cos k\theta.
\end{aligned}$$

So $a_1 + a_4 = 2\cos\theta$ and $a_2 + a_3 = 2\cos 2\theta$.

Putting these values into the elementary symmetric polynomial we have

$$2\cos\theta + 2\cos 2\theta = -1$$
$$\Rightarrow \quad 4\cos^2\theta + 2\cos\theta - 1 = 0$$
$$\Rightarrow \quad \alpha^2 + 2\alpha - 1 = 0.$$

Taking the positive root of this quadratic we have

$$\alpha = \frac{\sqrt{5}-1}{4}.$$

In the demonstrations in Section A.1.2 we showed how to construct the square root of a number so, with the construction of rational numbers and one square root, we can construct the angle $\theta = 2\pi/5$ to give the second vertex of the pentagon.

Gauss gave the following algebraic expression for constructing the 17-gon [ED, p32]:

$$\begin{aligned}\cos(2\pi/17) &= -\frac{1}{16} + \frac{1}{16}\sqrt{17} + \frac{1}{16}\sqrt{34-2\sqrt{17}} \\ &+ \frac{1}{8}\sqrt{17+3\sqrt{17}-\sqrt{34-2\sqrt{17}}-2\sqrt{34+2\sqrt{17}}}.\end{aligned}$$

This can be derived from the cyclotomic equation of degree 17 by forming equations of the roots similar to the one we used for the pentagon above although more complicated [ST, pp171–173]. Research into constructions of n-gons has continued and methods have been found for the 257-gon. There has been some research into the 65,537-gon, with Richelot publishing a paper as early as 1832, although Stewart is not clear on whether this resulted in a satisfactory construction [ST, p170]. It is difficult to imagine the complexity of solving trigonometric equations in the 65,536 primitive roots of cyclotomic equation, particularly without the advanced computer packages now available to us.

Bibliography

[AL] Allenby, R. B. J. T. *Rings, Fields and Groups: An Introduction to Abstract Algebra* Edward Arnold, London 1983

[DU] Durbin, John R. *Modern Algebra: An Introduction* (3rd Edition). John Wiley & Sons, New York, 1992

[ED] Edwards, Harold M. *Galois Theory* Springer-Verlag, New York, 1984.

[GA] Gaal, Lisl *Classical Galois Theory with Examples.* Markham Publishing, Chicago, 1971

[ST] Stewart, Ian *Galois Theory* (2nd edition) Chapman and Hall, London, 1989.

[WA] van der Waerden, B. L. *Algebra: Volume 1.* Translated by Fred Blum and John R Schulenberger. Frederick Ungar Publishing, New York, 1970.

Epilogue

The assessors for this project were unanimous in awarding this student a first. The overall standard of the mathematics was very high, and the student obviously understood the sometimes quite difficult material. This was obvious from both the oral presentation as well as at the viva-voce. *The style of the presentation is mature and the student's judgement on what to include and what to leave as a reference is nigh perfect. The layout is good (although this is less obvious here as it has had to change to fit the LATEX style of this text) and leaves the reader no doubt that the student had read and understood the books referred to.*

The only criticisms were the odd technical lapse and typographical error (many have been left in—maybe you can spot them).

The student took the responsibility to read and understand about Galois Theory. This is a commitment and involves being willing to devote a great deal of time to reading, re-reading and understanding books that can be quite advanced. It also meant doing many examples. Therefore, much time needs to be put aside which is consistent with the amount the project counts (one sixth of the final year in this instance). The self confidence of the student is also obvious from the writing, and this indicates a high level of mathematical ability.

B
Project Example 2: Algebraic Curves

Preamble

This projects takes further a topic with which mathematics students are familiar, namely graphs of functions. As in Appendix A, the subject is well embedded in the literature, but here the technicalities are less. There are places where the student has to branch out into the unknown (for the student), but much of the project is putting what is reasonably familiar into more rigorous mathematical terms.

Abstract

This project is an investigation into the theory of algebraic curves. We start by considering the basic definition and characteristics of an algebraic curve. We consider how an algebraic curve lies in a projective space and this leads on to studying the intersection of two curves. The properties of some simple curves and how they intersect is considered in more detail including the group structure of elliptic curves. We also investigate some results from complex analysis and subsequently use them to define Riemann surfaces and explain their connection with algebraic curves. This leads to more advanced theories from complex function theory such as integration on a curve. The topology of algebraic curves as subsets of $\mathbf{C}^2$ is also considered.

Introduction

Algebraic curves have a history of over 2000 years and it is inevitable that the Greeks made the first developments and discoveries. The Greeks had a limited algebraic knowledge but good geometric methods and to Euclid a circle was not defined by an equation but by the locus of all points having an equal distance from a fixed point. Greek mathematics was virtually ignored in Western Europe until it was rediscovered in the Renaissance period. The Arabs were responsible for reintroducing Greek mathematics into Western society and also brought with them a sophisticated algebraic knowledge and a good algebraic notation.

In the late seventeenth century a more algebraic approach was being applied to algebraic curves and a large number of mathematicians published papers on the subject. These included Newton who made a study of cubic curves and identified and classified 72 different cases. By the nineteenth century it was soon apparent that considering curves as subsets of $\mathbf{C}^2$ was easier than looking at curves as subsets of $\mathbf{R}^2$ as a curve in the Euclidean plane can degenerate so much that it no longer resembles a curve.

In the nineteenth century mathematicians such as Riemann were interested in more abstract geometrical spaces and this led to his theory of Riemann surfaces and their connection with algebraic curves. Research into algebraic curves continues up to the present day particularly in the area of particle physics. Here the 'lifetime' of a particle is represented by a 'string' in four dimensional space-time. This can consequently be represented by a Riemann surface.

Algebraic curves is one of the most varied and beautiful subjects in mathematics. This project does not have a specific aim and the discussion is relatively informal, but it does give an insight into some of the more interesting and appealing results that have arisen. As one can imagine, for a subject that has existed for such a long time, the theory of algebraic curves is vast and involves a lot of interplay from different areas of mathematics such as algebra, topology, complex analysis and geometry. It would therefore be foolish to try and incorporate every idea in the subject: consequently a lot of very elegant and compelling theories have been missed out; these include the Riemann–Roch theorem and an in-depth investigation into singularities.

In Section B.1 we show that an algebraic curve is defined by a polynomial and look at why we consider curves in $\mathbf{C}^2$. We also define singularities and consider the multiplicity of such points. In Section B.2 we investigate how two curves intersect by considering one of the most famous theorems in the study of algebraic curves: Bezout's theorem. Section B.3 applies results from previous sections to some simple curves namely conics and cubics. We shall see that a cubic curve gives rise to some amazing properties such as the additive group structure. Section B.4 is a resumé of results from complex analysis that we need to be familiar with for the final chapters.

In Section B.5 we investigate the topology of curves in $\mathbf{C}^2$ which is a space that has four real dimensions. We find that these objects are topological surfaces and we consequently introduce the idea of an abstract Riemann surface to represent an algebraic curve. Section B.6 is a less rigid investigation into more advanced ideas involving algebraic curves and complex analysis which hence leads on to topics from complex function theory.

Throughout the course of the project three books were heavily relied upon for the theorems and subsequent proofs. These were Frances Kirwan *Complex Algebraic Curves*, Bak and Newman *Complex Analysis* and R. J. Walker, *Algebraic Curves* and particularly Bak and Newman in Section B.4 and Kirwan in Sections B.5 and B.6.

B.1 Basic Definitions and Properties

In this section we will look at some basic properties of algebraic curves. We also investigate the concept of a projective space and why it is useful to consider curves in the projective plane as well as curves in $\mathbf{C}^2$.

B.1.1 Complex Algebraic Curves and Real Algebraic Curves

A complex algebraic curve C in $\mathbf{C}^2$ is defined by

$$C = \{(x,y) \in \mathbf{C}^2 : P(x,y) = 0\}$$

where $P(x,y)$ is a polynomial in two variables with complex coefficients. All complex algebraic curves are subsets of $\mathbf{R}^2$ namely

$$C = \{(x,y) \in \mathbf{R}^2 : P(x,y) = 0\}$$

where $P(x,y)$ has real coefficients.

For every real algebraic curve there is an associated complex algebraic curve defined by the same polynomial. Investigating curves in the complex space is far simpler than only considering curves in the Euclidean space $\mathbf{R}^2$. If we work with complex numbers every polynomial factorises completely in $\mathbf{C}$ (not the case in $\mathbf{R}$). This result from complex analysis, known as the fundamental theorem of algebra (see B.4.31), allows us to completely factorise any polynomial, so for simplicity in the remainder of the project we will only usually consider curves as subsets of $\mathbf{C}^2$.

The *degree* of a curve C is given by the degree of the polynomial that defines C. Hence the curve defined by $P(x,y) = x(x^2+1)$ has degree 3.

The polynomial $P(x_1,\ldots,x_n) = 0$ is *homogeneous* of degree d if and only if

$$P(tx_1,\ldots,tx_n) = t^d(x_1,\ldots,x_n)$$

for all $t, x_1, \ldots, x_n$ and some d, i.e. P is homogeneous if every one of its terms are of the same degree d.

A polynomial is *irreducible* if it has no factors apart from constants and scalar multiples of itself. A curve defined by $P(x,y)$ is irreducible if $P(x,y)$ is irreducible.

So far we have only considered curves in $\mathbf{C}^2$, but a curve in $\mathbf{C}^2$ is never compact as it always has branches going to infinity. We can compactify a curve in $\mathbf{C}^2$ by adding in the "points at infinity" and getting a projective space. It can be beneficial to consider curves in a projective space, for example consider two parallel lines which lie in $\mathbf{C}^2$. They will of course never intersect but if we add a point at infinity to each line they meet at a distinct point (albeit at "infinity" for this example). Consequently any two lines in a projective space will meet at a distinct point. For this reason it is useful to consider curves in a projective space studying their algebraic properties.

B.1.2 Projective Spaces

B.1.2.1 Complex Projective Plane. Consider the set of all possible complex lines which pass through the origin of $\mathbf{C}^3$. It can be shown that the set of all such lines form a complex surface (4 real dimensions). The surface that is formed is known as the *complex projective plane* and denoted by $\mathbf{CP}_2$. This idea can be generalised to give a projective space for any complex dimension n.

The complex projective space $\mathbf{CP}_n$ of dimension n is the set of complex one dimensional subspaces of the vector space $\mathbf{C}^{n+1}$. When $n = 2$ we get the complex projective plane $\mathbf{CP}_2$ as above.

To prove results algebraically we need a notation to specify points in $\mathbf{CP}_2$. Clearly a line through the origin of $\mathbf{C}^3$ is uniquely determined if we choose any point (save the origin) which lies on that line. So a point p in $\mathbf{CP}_2$ consists of the unique line in $\mathbf{C}^3$ which passes through $(0,0,0)$ and (a,b,c) where a, b and c are not all zero. The notation for such a point p is given by $[a,b,c]$.

We can define the points in $\mathbf{CP}_2$ by

$$\mathbf{CP}_2 = \{[a,b,c] : (a,b,c) \in \mathbf{C}^3 - \{0\}\}$$

and also $[a, b, c] = [d, e, f]$ if and only if $a = \lambda d$, $b = \lambda e$ and $c = \lambda f$ for some $\lambda \in \mathbf{C} - \{0\}$. So far we have used the theory of three dimensional complex space to develop the idea of a projective space. However we want to use this space to study objects which have two complex dimensions, specifically algebraic curves. So we must find a way of associating figures in $\mathbf{CP}_2$ with figures in $\mathbf{C}^2$.

Suppose we have a projective figure which lies in $\mathbf{CP}_2$. We can place a plane in $\mathbf{C}^3$ making sure it does not pass through the origin. The corresponding figure in $\mathbf{C}^2$ are the points of the figure that pierces the plane. This is okay provided that each of the projective figures pierces the plane, but any point of $\mathbf{CP}_2$ which consists of a line through the origin parallel to the plane that does not pierce the plane. Such a point is called an *ideal point* for that plane. In general a projective figure can be represented by the points which pierce a plane in $\mathbf{C}^3$ together with some or all the ideal points for that plane. Using this idea we could have defined the complex projective plane as simply the usual complex plane with "ideal points at infinity", one ideal point for each complex line through the origin. In the following example we will show how points in $\mathbf{C}^2$ can be identified with points in $\mathbf{CP}_2$.

Example 1.22 Consider the subset S of points in $\mathbf{CP}_2$ which pierce the plane $z = 1$:
$$S = \{[x, y, z] : z \neq 0\}.$$
Each point $(x, y) \in \mathbf{C}^2$ can be identified with the point $(x, y, 1) \in S$. Conversely, $[x, y, z] = [x/z, y/z, 1] \in S$ can be identified with $(x/z, y/z) \in \mathbf{C}^2$. This gives a 1–1 correspondence between $\mathbf{C}^2$ and S. So $\mathbf{C}^2$ can be regarded as a subset of $\mathbf{CP}_2$.

B.1.2.2 Projective Transformations. A geometry consists of a space together with a group of transformations that act on that space: projective geometry is no different. Since the points of $\mathbf{CP}_2$ consist of lines through the origin of $\mathbf{C}^3$ a projective transformation maps straight lines on to straight lines. Consequently transformations in $\mathbf{CP}_2$ must be linear transformations of $\mathbf{C}^3$. If we define the map $\Phi: \mathbf{C}^3 - \{0\} \to \mathbf{CP}_2$ by $\Phi(a, b, c) = [a, b, c]$ then we can define a *hyperplane* in $\mathbf{CP}_2$ as the image of $V - \{0\}$ under Φ where V is a two-dimensional complex subspace of $\mathbf{C}^3$.

Proposition 1.24 Given four distinct points p_0, p_1, p_2 and q of $\mathbf{CP}_2$ no three of which lie on a hyperplane, there is a projective transformation which takes P_0 to $[1, 0, 0]$, p_1 to $[0, 1, 0]$, p_2 to $[0, 0, 1]$ and q to $[1, 1, 1]$.

Proof Let u_0, u_1, u_2 and v be elements of $\mathbf{C} - \{0\}$ whose image under Φ are

p_0, p_1, p_2 and q. So u_0, u_1, u_2 form a basis of $\mathbf{C}^3$. There is therefore a unique linear transformation α of $\mathbf{C}^3$ taking u_0, u_1, u_2 to the standard basis $(1,0,0)$, $(0,1,0)$, $(0,0,1)$. Now $\alpha(v) = (\lambda_0, \lambda_1, \lambda_2)$ where $\lambda_0, \lambda_1, \lambda_2$ are non-zero complex numbers.

Hence the composition of α with a 3×3 matrix

$$\begin{bmatrix} 1/\lambda_0 & 0 & 0 \\ 0 & 1/\lambda_1 & 0 \\ 0 & 0 & 1/\lambda_2 \end{bmatrix}$$

defines a projective transformation taking p_0 to $[1/\lambda_1, 0, 0] = [1, 0, 0]$, p_1 to $[0, 1/\lambda_2, 0] = [0, 1, 0]$, p_2 to $[0, 0, 1/\lambda_3] = [0, 0, 1]$ and q to $[1, 1, 1]$. So when we are investigating a certain point on a curve we can assume that under a suitable projective transformation that the point can be a point of reference, namely $[1, 0, 0]$, $[0, 1, 0]$, or $[0, 0, 1]$.

B.1.3 Affine and Projective Curves

Now we can define a *projective curve* in $\mathbf{CP}_2$ by a homogeneous polynomial $P(x, y, z)$ in x, y and z. Thus a projective curve C is defined by

$$C = \{[x, y, z] \in \mathbf{CP}_2 : P(x, y, z) = 0\}.$$

Algebraic curves in $\mathbf{C}^2$ as defined in Section B.1.1 will be called *affine curves* to distinguish them from projective curves. Affine and projective curves are closely related and it was shown that points in $\mathbf{C}^2$ can be identified with points in $\mathbf{CP}_2$. This idea can be extended for affine and projective curves.

Let $P(x, y, z) = 0$ be a curve in $\mathbf{CP}_2$ which does not have $z = 0$ as a component. The curve has an associated non-homogeneous polynomial $P(x, y, 1)$ of the same degree. Consequently an affine curve $P(x, y)$ in $\mathbf{C}^2$ can be identified with $z^d P(x/z, y/z)$ in $\mathbf{CP}_2$ where d is the degree of the polynomial.

B.1.4 Singular Points

A singular point is basically a point where the curve does not look "smooth". This idea can be best illustrated by looking at the sketches of algebraic curves in $\mathbf{R}^2$ given at the end of this section.

Mathematically, for an affine curve C a *singular point* or *singularity* is where

$$\frac{\partial P}{\partial x}(a, b) = 0 = \frac{\partial P}{\partial y}(a, b)$$

where $P(x, y) = 0$ defines C and $(a, b) \in C$.

Similarly for a projective curve, we have

$$\frac{\partial P}{\partial x}(a, b, c) = \frac{\partial P}{\partial y}(a, b, c) = \frac{\partial P}{\partial z}(a, b, c) = 0, \quad \text{where} \quad (a, b, c) \in C.$$

If a curve does not have any singularities it is termed *non-singular* i.e. $x^2 + y^2 - 1 = 0$. The *multiplicity* of an affine curve defined by $P(x, y)$ at a point (a, b) is the smallest possible integer m such that

$$\frac{\partial^m P}{\partial x^i \partial y^j}(a, b) \neq 0 \quad \text{for some } i \leq 0, j \leq 0 \text{ where } i + j = m.$$

Similarly for a projective curve the multiplicity of a point (a, b, c) is the smallest possible integer m such that

$$\frac{\partial^m P}{\partial x^i \partial y^j \partial z^k}(a, b, c) \neq 0 \quad \text{where } i + j + k = m.$$

The point is non-singular if its multiplicity is one. If this is the case it has a unique tangent line at (a, b). A point is a double point, triple points, etc. A singular point is *ordinary* if the first non-vanishing terms expanded about this point have no repeated factors, it thus has m distinct tangent lines at (a, b). We will finish this section by looking at some types of singular points in $\mathbf{R}^2$.

Example 1.41

$$P(x, y) = x^3 - x^2 + y^2 \quad \text{see Figure B.1.}$$

The origin is a double point with two distinct tangents. This type of double point is called a node.

Example 1.42

$$P(x, y) = x^3 - y^2 \quad \text{see Figure B.2.}$$

This double point is not ordinary as it has only one tangent line that runs along the x-axis. This type of singularity is known as a cusp. **Example 1.43**

$$P(x, y) = (x^4 + y^4)^2 - x^2 y^2 \quad \text{see Figure B.3.}$$

This is not an ordinary point of multiplicity four as it has two tangents at right angles running along the x and y axis.

Example 1.44

$$P(x, y) = (x^4 + y^4 - x^2 - y^2)^2 - 9x^2 y^2 \quad \text{see Figure B.4.}$$

Here we have an ordinary point of multiplicity four with four distinct tangents which pass through the origin.

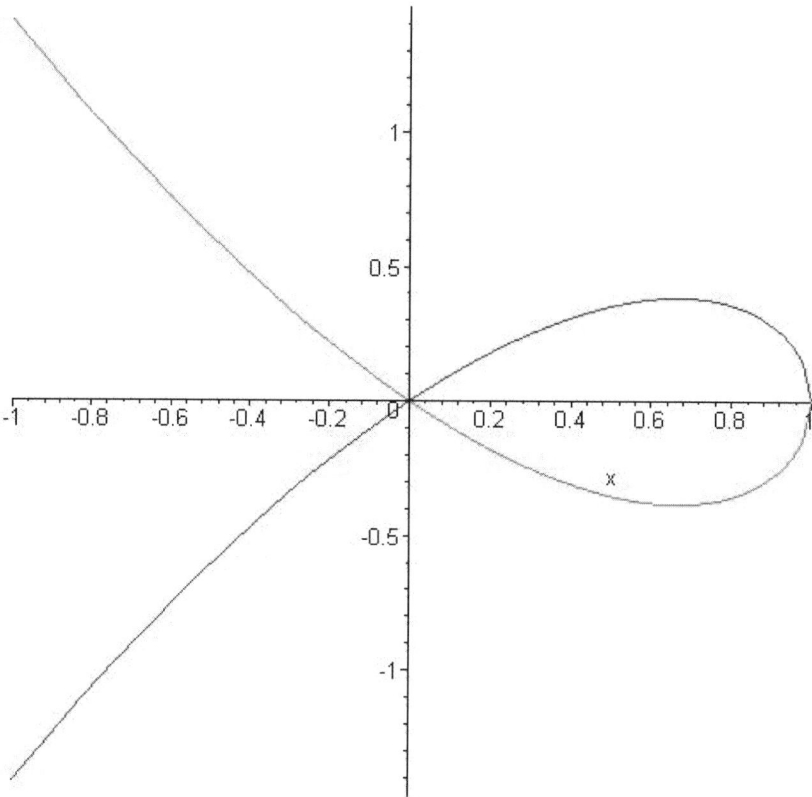

Figure B.1   The curve $x^3 - x^2 + y^2 = 0$

B.2 Intersection of Two Curves and Points of Inflection

This section investigates whether two polynomials in x, y and z have any common solutions, i.e. how do two projective curves intersect? We shall also look at points of inflection on a curve.

B.2.1 Bezout's Theorem

The number of points of intersection between two projective curves defined by the polynomials $P(x, y, z)$ and $Q(x, y, z)$ is given by Bezout's Theorem. If P is

B. Project Example 2: Algebraic Curves

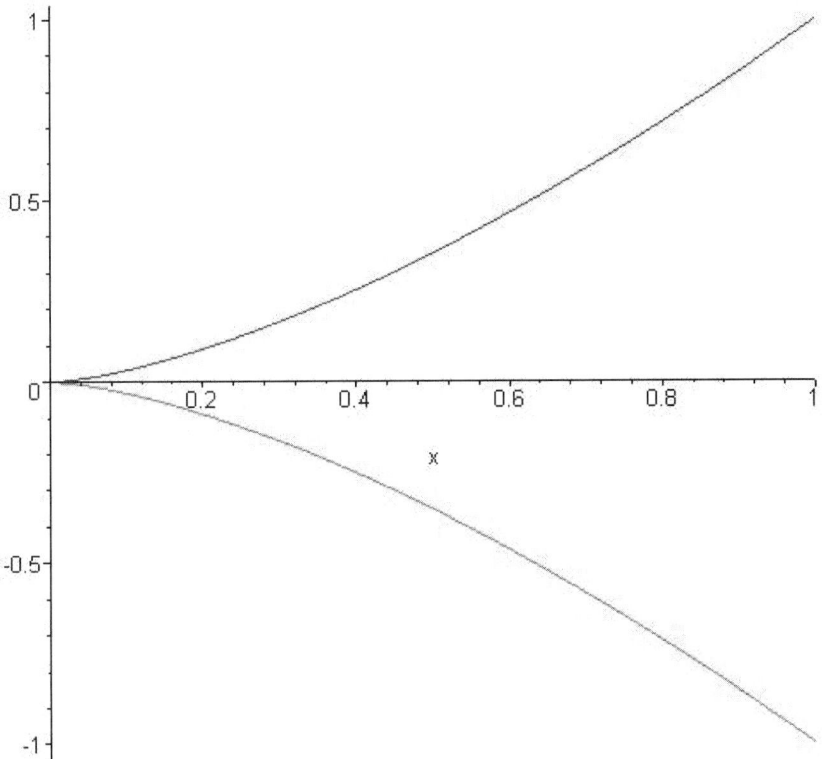

Figure B.2 The curve $x^3 - y^2 = 0$

of degree n and Q is of degree m then P and Q have precisely nm common solutions. To prove this we first need the following algebraic result.

B.2.1.1 Resultant of Two Polynomials. We can rearrange P and Q so that

$$P(x,y,z) = a_0(y,z) + a_1(y,z)x + a_2(y,z)x^2 + \ldots + a_n(y,z)x^n$$

and

$$Q(x,y,z) = b_0(y,z) + b_1(y,z)x + b_2(y,z)x^2 + \ldots + b_m(y,z)x^m$$

where $a_i, b_i \in \mathbf{C}$.

The *resultant* $R(y,z)$ of P and Q is given by the determinant of

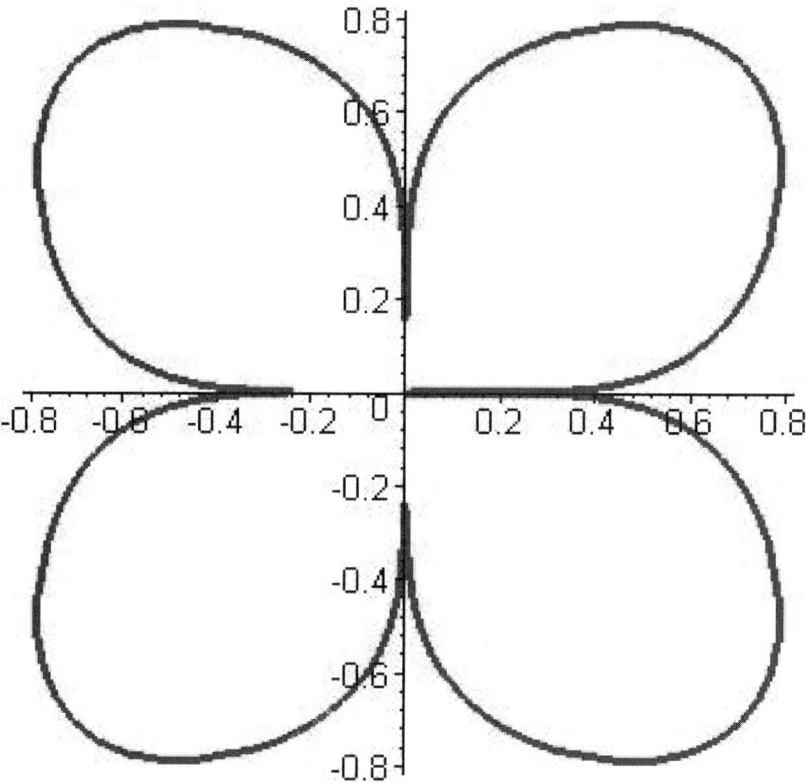

Figure B.3   The curve $(x^4+y^4)^2 - x^2y^2 = 0$

$$\begin{bmatrix} a_0(y,z) & a_1(y,z) & \ldots & a_n(y,z) & 0 & 0 & \ldots \ldots & 0 \\ 0 & a_0(y,z) & a_1(y,z) & \ldots & a_n(y,z) & 0 & \ldots \ldots & 0 \\ \vdots & \vdots & \vdots & \vdots & \vdots & \vdots & \vdots \vdots & \vdots \\ 0 & 0 & \ldots & 0 & a_0(y,z) & a_1(y,z) & \ldots \ldots & a_n(y,z) \\ b_0(y,z) & b_1(y,z) & \ldots & \ldots & \ldots & b_m(y,z) & 0 \ldots & 0 \\ \vdots & \vdots & \vdots & \vdots & \vdots & \vdots & \vdots \vdots & \vdots \\ 0 & \ldots & 0 & b_0(y,z) & b_1(y,z) & \ldots & \ldots \ldots & b_m(y,z) \end{bmatrix}$$

$R(y,z)$ is thus a polynomial in y and z of degree nm.

If $P(x,y,z)$ and $Q(x,y,z)$ are non-constant homogeneous polynomials we can arrange it so that
$$P(1,0,0) \neq 0 \neq Q(1,0,0)$$
which ensures that they have the same degree when regarded as polynomials in x with coefficients in y and z. It follows that $P(x,y,z)$ and $Q(x,y,z)$ have

B. Project Example 2: Algebraic Curves

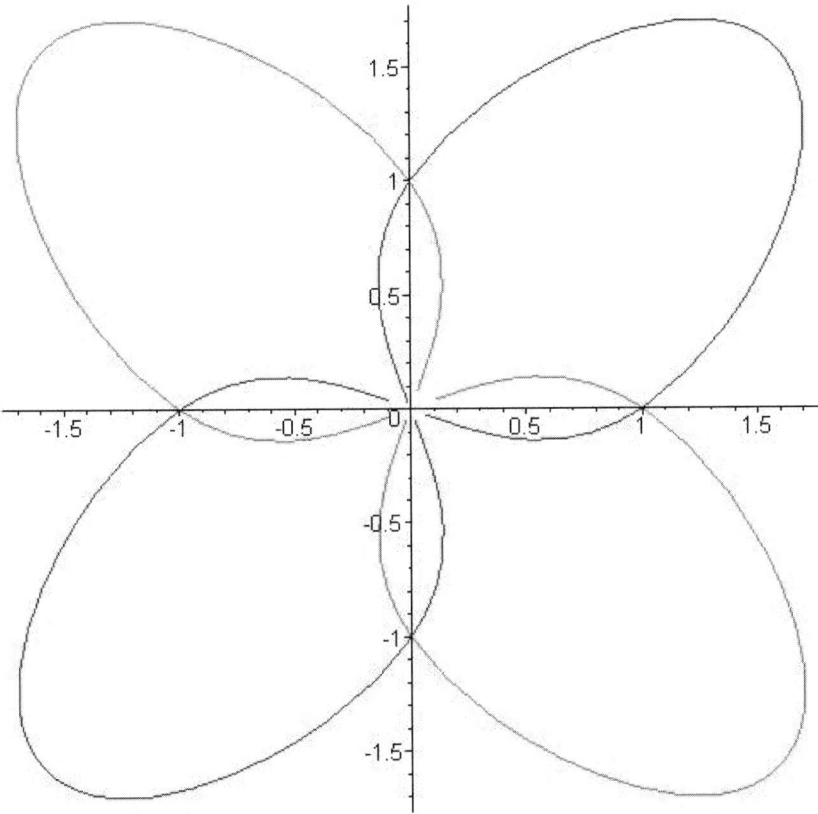

Figure B.4   The curve $(x^4 + y^4 - x^2 - y^2)^2 - 9x^2y^2 = 0$

a non-constant common factor if and only if $R(y, z) = 0$ and it always can be arranged so that this condition is true. Using this result we can now prove the weak form of Bezout's theorem

Theorem 2.12

If two projective curves of degrees n and m have more than nm common points, then they have a common component.

Proof [Walker 50, 3.1 p59]

Let C and D be two projective curves of degrees n and m defined by the polynomials $P(x,y,z)$ and $Q(x,y,z)$ that have more than nm common points. We can select any set of $nm+1$ of these points and join each selected pair by a line. As there is only a finite set of such lines then there is a point p not on any of the lines or on C or D. Under a suitable projective transformations we can choose this point to be $p = [1,0,0]$. Consequently P and Q can be put in the form

$$P(x,y,z) = a_0(y,z) + a_1(y,z)x + a_2(y,z)x^2 + \ldots + a_n(y,z)x^n$$

and

$$Q(x,y,z) = b_0(y,z) + b_1(y,z)x + b_2(y,z)x^2 + \ldots + b_m(y,z)x^m$$

where $a_i, b_j \in \mathbf{C}$. The resultant $R(y,z)$ of P and Q is either zero or a homogeneous polynomial of degree nm in y and z. We deduce that $R(b,c) = 0$ if and only if there is an a such that $P(a,b,c) = 0 = Q(a,b,c)$, that is to say the coordinates b,c common to P and Q satisfy $R(y,z) = 0$. However each of the $nm+1$ points has a different value for the ratio $b:c$ since no pair of them are collinear with $[1,0,0]$. Hence $R(y,z) = 0$ and so C and D have a common factor, i.e. two curves intersect at nm points at most.

In section B.1 we defined the multiplicity of a curve. We use a similar idea to define the intersection multiplicity of two curves and use it to show that the number of intersections counted properly is precisely nm, the strong case of Bezout's theorem.

Theorem 2.13

If C and D are two projective curves of degrees n and m in $\mathbf{CP}_2$ which have no common component then they have precisely nm points of intersection counted properly; i.e.

$$\sum_{p \in C \cap D} I_p(C,D) = nm.$$

The intersection multiplicity can be defined by using the resultant of two polynomials as defined in Section B.2.1. It follows that from our definition of the resultant that the intersection multiplicity $I_p(C,D)$ for two projective curves C and D at a point p has the following properties:

* $I_p(C,D) = I_p(D,C)$.

* $I_p(C,D) = \infty$ if the point p lies on a common component of C and D otherwise $I_p(C,D)$ is a non-negative integer.

* $I_p(C,D)$ if and only if the point p does not lie on $C \cap D$.

* Two distinct lines meet with intersection multiplicity one.

B. Project Example 2: Algebraic Curves

* If C_1 and C_2 are defined by the homogeneous polynomials $P(x,y,z)$ and $Q(x,y,z)$ and C is defined by $R(x,y,z) = P(x,y,z)Q(x,y,z)$ then $I_p(C,D) = I_p(C_1,D) + I_p(C_2,D)$.

* If C and D are defined by $P(x,y,z)$ and $Q(x,y,z)$ of degrees n and m and E is defined by $PQ + R$ where $R(x,y,z)$ is homogeneous of degree $m - n$ then $I_p(C,D) = I_p(C,E)$.

From this we can define the intersection multiplicity $I_p(C,D)$ for some $p = C \cap D$ as the largest k where $(bz - cy)^k$ divides the resultant $R(y,z)$ and it can be shown that k is uniquely determined. Using this definition we can prove the strong case of Bezout's theorem.

Proof of Theorem 2.13 [Kirwan 92, 3.1 p62] Under a projective transformation we can choose a point $p = [1,0,0]$ such that $p \notin C \cup D$. p does not lie on a line containing two distinct points of $C \cap D$ and p does not lie on a tangent line to C or D at any point of $C \cap D$. If C and D are defined by the homogeneous polynomials $P(x,y,z)$ and $Q(x,y,z)$ then the resultant $R(y,z)$ is a homogeneous polynomial of degree nm in y and z.

The resultant $R(y,z)$ can be expressed as a product of linear factors

$$R(y,z) = \prod_{i=1}^{k}(c_i z - b_i y)^{e_i} \qquad \text{where} \qquad c_i, e_i \in \mathbf{C}$$

and $e_i = nm = e_1 + e_2 + \ldots + e_k$.

There exists a unique complex number a_i such that $C \cap D = \{p_i : i \leq i \leq k\}$ where $p_i = [a_i, b_i, c_i]$ which leads to $I_{p_i}(C,D) = e_i$ and hence

$$\sum_{p \in C \cap D} I_p(C,D) = nm$$

proving the result.

□

B.2.2 Points of Inflection on a Curve

A point p on a curve C in $\mathbf{CP}_2$ is *a point of inflection* of C when the tangent to C at p is not a component of C and meets C at p with a multiplicity of at least three. We want a way to compute such points: for this we use the *Hessian* curve. If C is defined by the homogeneous polynomial $P(x,y,z)$ then the Hessian $H(x, y, z)$ of C is given by the determinant of the 3×3 matrix

$$\begin{bmatrix} P_{xx} & P_{xy} & P_{xz} \\ P_{yx} & P_{yy} & P_{yz} \\ P_{zx} & P_{zy} & P_{zz} \end{bmatrix}.$$

When H vanishes at a point of C then that point is an inflection point of C, for example if $[a, b, c]$ lies on C and $H(a, b, c) = 0$ then $[a, b, c]$ is a point of inflection (or flex) of C. The second derivatives of $P(x, y, z)$ are of degree $d-2$. It follows that H must be a polynomial of degree $3(d-2)$ in x, y and z. We can easily deduce from this and from Bezout's theorem that a non-singular curve of degree greater than or equal to three has at least one point of inflection, i.e. when $d = 2$, H is a polynomial of degree zero which no a, b, c could satisfy. When $H = 0$ or a constant then the curve that leads to this is defined as having no Hessian.

B.3 Conics and Cubics

To illustrate some of the results that were investigated in Section B.2 it would be useful to apply them to some simple curves. It can be easily seen that curves of order one, namely lines intersect at only one point and that every point on the line is a point of inflection, that is to say its second derivative vanishes.

B.3.1 Conics

A *conic* is a curve of degree 2 in $\mathbf{CP}_2$. Any non-singular projective conic C in $\mathbf{CP}_2$ is equivalent under a projective transformation to the conic

$$x^2 = yz.$$

B.3.2 Cubics

Cubics are curves of degree 3 in $\mathbf{CP}_2$. Cubics are the first class of curve where some more interesting properties occur. Firstly we will show that a non-singular cubic can be written in the standard form $y^2 = g(x)$.

B. Project Example 2: Algebraic Curves

Theorem 3.21

By a proper choice of coordinates any non-singular cubic C can be put in the form
$$y^2 = g(x) \qquad (1)$$
where $g(x)$ is a cubic polynomial with distinct roots.

Proof[Walker 50, p72]

To prove this we need the result from Section B.2 that every non-singular curve of degree $leq 3$ has at least one point of inflection. Under a projective transformation we can choose C to have a point of inflection at $[0, 0, 1]$, which reduces (1) to
$$x^3 + h(x, y) = 0 \qquad (2)$$
where $h(x, y)$ is of degree 2. Also $h(x, y)$ must include a term ay^2 (a not zero); if not $[0, 0, 1]$ would be a singularity. Solving (2) for y we get
$$y = \alpha x + \beta + \sqrt{g(x)}.$$
If we make the linear transformation $y' = y - \alpha x - \beta$ and $x' = x$ we obtain
$$y' + \alpha x' + \beta = \alpha x' + \beta + \sqrt{g(x)}.$$
Squaring and dropping the dashes we get the required form $y^2 = g(x)$.

$\square$

If we had a singular cubic then $g(x)$ would not have distinct roots. If this were the case under a suitable projective transformation $g(x) = x^2(x+1)$ for a nodal cubic or $g(x) = x^3$ for a cuspidal cubic. We can now investigate one of the most famous properties of cubics.

B.3.2.1 Inflections on a Cubic. We have the following theorem,

Theorem 3.23

A non-singular cubic C has 9 distinct points of inflection with the property every line joining two of them contains a third.

Proof[Walker 50, 6.6, p 72]

Under the linear transformation $x' = ax + b$ and $y' = y$, $y^2 = g(x)$ can be written
$$F = y^2 - x^3 - ax^2 - bx = 0$$
with $b(a^2 - 4b) \neq 0$ since $x^3 + ax^2 + bx = 0$ has distinct roots. We first look for points in the finite plane. The Hessian of F is given by
$$H = (y^2 + bx)(3x + a) - (ax + b)^2.$$
Eliminating y between F and H gives
$$p(x) = 3x^4 + 4ax^3 + 6bx^2 - b^2 = 0.$$
So $p(x)$ has 4 distinct roots and $p'(x) = 12(x^3 + ax^2 + bx)$ and the resultant of p and p' is
$$b^4(a^2 - 4b)^2 \neq 0.$$
So for each x value there are two y values. None of these values of x satisfy $x^3 + ax^2 + bx = 0$. So this makes eight points of inflection plus a flex at infinity (i.e. at $[0, 0, 1]$ from the assumption made in Theorem 3.21) which makes a total of nine. If we choose two points of inflection in the finite plane to be $[1, a, b]$ and $[1, a, -b]$ say, then there must be a line that passes through these two points. If then we take the flex at infinity namely $[0, 0, 1]$ then a line which contains the first two points must also contain this point. We can arrange the nine points of inflection into an array such that three points in any row, column or diagonal lies on a line in $\mathbf{CP}_2$. We now look at the result that connects conics, cubics, and Bezout's Theorem. Let C and D be projective cubics defined by the polynomials $P(x, y, z)$ and $Q(x, y, z)$ and assume C and D meet at exactly nine points $p_1, \ldots, p_9$. Now let E be another projective cubic defined by $R(x, y, z)$ where E contains the points $p_1, \ldots, p_8$. If C, D and E are linearly independent then there is a curve $C(\lambda, \mu, \nu)$ defined by
$$\lambda P(x, y, z) + \mu Q(x, y, z) + \nu R(x, y, z) = 0 \quad \text{for some } \lambda, \mu, \nu$$
which passes through the points $p_1, \ldots, p_8$ and any two arbitrary points q and r say. We will show that this in turn leads to a contradiction and therefore if E contains $p_1, \ldots, p_8$ then E contains p_9. By Bezout's Theorem, no line can contain four of these points as such a line would be a common component of C and D. Similarly no cubic can contain seven of these points. Now if p_6, p_7 and p_8 lie on a line L then $p_1 \ldots, p_5$ lie on a unique conic Q. This conic is unique because if two conics lie on five common points they must have a common component (Bezout's Theorem), the remaining components can only have one intersection not on this line, and hence the other four points must be collinear.

B. Project Example 2: Algebraic Curves   211

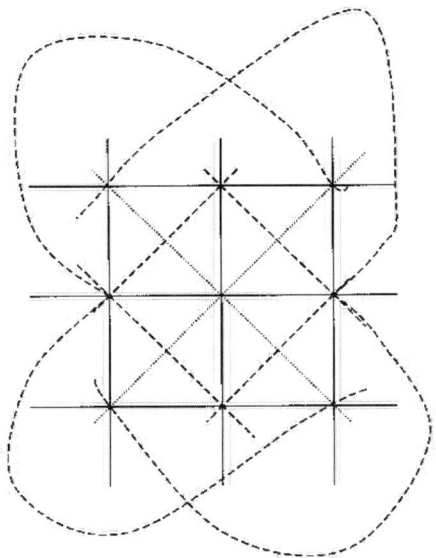

Figure B.5   The array in $\mathbf{CP}_2$

Now let $q \in L$ and $r \notin L \cup Q$. $C(\lambda, \mu, \nu)$ contains $p_1, \ldots, p_8$ and λ, μ, ν are chosen so that it contains q and r. So $C(\lambda, \mu, \nu)$ must have L as a component, the other component is therefore Q. Hence $C(\lambda, \mu, \nu) = L \cup Q$, but this is impossible since $r \notin L \cup Q$. Now if $p_1, \ldots, p_6$ lie on a conic Q then p_7, p_8 lie on a line L. If we choose $C(\lambda, \mu, \nu)$ to vanish at the points $q \in Q$ and $r \notin L \cup Q$ then this again leads to a contradiction since $C(\lambda, \mu, \nu) = L \cup Q$. Finally if no three of $p_1, \ldots, p_8$ lie on a line and no six on a conic, we let L be the line where p_1, p_2 lie and Q be the unique conic on which $p_3, \ldots, p_7$ lie. If we take $q \in L$ and $r \in L$ then this again leads to a contradiction since $p_8 \notin Q$ and

$C(\lambda, \mu, \nu) = L \cup Q$ as before. This exhausts all possibilities so we can conclude that the curve $C(\lambda, \mu, \nu)$ does not pass through $p_1, \ldots, p_8, q$ and r for any choice of q and r. So E is linearly dependent on C and D, and hence E must contain p_9 as well. We can use this result in the next section.

$\square$

B.3.3 Additive Group Structure on a Cubic

Theorem 3.31

Given any non-singular projective cubic C in $\mathbf{CP}_2$ and a point of inflection p_0 on C there is a unique additive group structure on C such that p_0 is the zero element and three points of C add up to zero if and only if the intersection of C with some line in $\mathbf{CP}_2$ (allowing for some multiplicities).

Proof [Kirwan 92 3.38, p 77]

Firstly we have to show that additions and inverse operations are uniquely determined. Now $-p_0 = p_0$ since p_0 is a point of inflection and if $p \neq p_0$ then $-p$ is the third point of intersection with some line in $\mathbf{CP}_2$ through p and p_0. So additive inverses are uniquely determined. Now let p and q be any points of C then $p + q = -r$ where r is the third point of intersection of C with a line in $\mathbf{CP}_2$ through p and q (if $p \neq q$) or the tangent to C at p (if $p = q$). So the addition operation is uniquely determined.

Now we have to show that there is an additive group structure on C with p_0 as the zero element. Commutativity, namely $p + q = q + p$, follows from the definition of the addition operation. Similarly $p + (-p) = p_0$ follows from the definition of additive inverses. We have to deduce that p_0 is the zero element.

For any $p \in C$ such that $p \neq p_0$ we have $p + p_0 = -r$ where r is the third point of intersection of C with the line in $\mathbf{CP}_2$ through p and p_0. This point r is not p_0, so $-r$ is the third point of intersection of C with the line in $\mathbf{CP}_2$ through r and p_0, which is the point p. Thus, $p + p_0 = p$ if $p \neq p_0$ and $p + p_0 = p_0$ since p_0 is a point of inflection, so p_0 is the zero element.

To prove associativity we let $p, q, r \in C$ and L_1 is the line in $\mathbf{CP}_2$ that meets C at $p, q, -(p+q)$ and similarly we let the following be lines in $\mathbf{CP}_2$ that meet C at

$$L_2 : p_0, p+q, -(p+q), \ L_3 : r, p+q, -((p+q)+r), \ M_1 : q, r, -(q+r),$$

$$M_2 : p_0, q+r, -(q+r) \text{ and } M_3 : p, q+r, -(p+(q+r)).$$

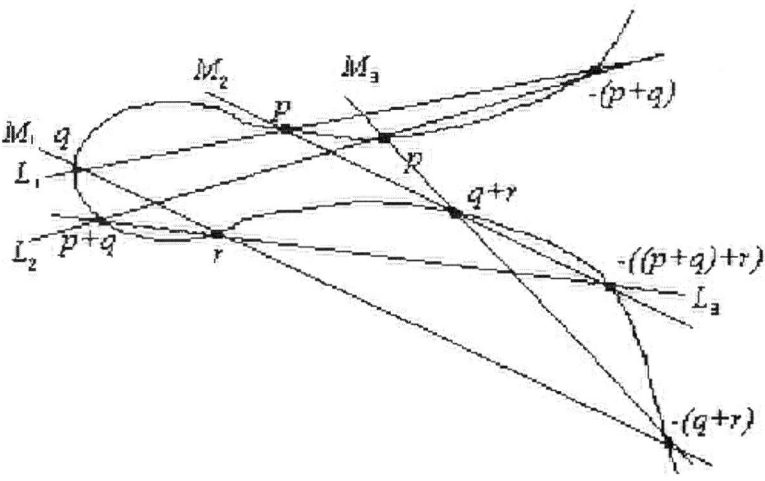

Figure B.6

Now let D and E be the reducible curves $D = L_1 \cup M_2 \cup L_3$ and $E = M_1 \cup L_2 \cup M_3$. C meets L_1 at $[p, q, -(p+q)]$ and therefore D meets C at

$$p_0, p, q, r, p+q, q+r, -(p+q), -(q+r), -((p+q)+r)$$

and by a similar deduction we find that E meets C at

$$p_0, p, q, r, p+q, q+r, -(p+q), -(q+r), -(p+(q+r)).$$

We have to show that $(p+q)+r = p+(q+r)$. We can see that C and D meet at nine points and that E contains eight of them. From the argument previously in the section, if two projective curves C and D intersect at nine points and if a projective curve E contains eight of them, then it also must contain the ninth. So $-(p+(q+r))$ and $-((p+q)+r)$ must describe the same point, hence $(p+q)+r = p+(q+r)$ proving associativity. Note that we used p_0 as the zero element.

In fact we could have used any point of C as long as we constructed our group around that point using the proviso that any three points of intersection with C add up to zero if they are three points of intersection of C with $\mathbf{CP}_2$.

□

B.4 Complex Analysis

For the final sections we need some standard results from complex analysis. We consider the definition of holomorphic functions and their properties and Cauchy's closed curve theorem. Although some proofs are given not every result is proved rigorously and the reader may refer to any introductory book on complex analysis for the details, e.g. [Bak and Newman 82].

We will use the notation $z = x + iy$ to represent a complex number and $f(z)$ denotes a complex function. $D(z_0 : r)$ is the open disc of radius $r > 0$ centred at z_0 (i.e. D is an open subset of C, or alternatively all neighbourhoods of z_0 contain such a disc.)

B.4.1 Holomorphic Functions and Entire Functions

A function f is only differentiable at z if $\dfrac{\partial f}{\partial x}$ and $\dfrac{\partial f}{\partial y}$ are continuous at z and it satisfies the Cauchy–Riemann equations:

$$\frac{\partial f}{\partial x} = i\frac{\partial f}{\partial y}.$$

Consequently it can be shown that $\dfrac{df}{dz} = \dfrac{\partial f}{\partial x}$. If D is an open subset of $\mathbf{C}$ and a complex number f is differentiable at every $z \in D$ then the function is *holomorphic* in D. If the first derivatives exist then it can be shown that all subsequent derivatives exist. The term *analytic* is often used to describe a holomorphic function.

Example 4.11

The function $f(x + iy) = x^2 - y^2 + 2ixy$ is holomorphic, but the function $f(x + iy) = x^2 - y^2 - 2ixy$ is not holomorphic.

Functions such as polynomials that are everywhere differentiable are called *entire* functions. An entire function is therefore infinitely differentiable.

B.4.2 Closed Curve Theorem and Line Integrals

B.4.2.1 Line Integrals. If we let $f(t) = u(t) + iv(t)$ be any continuous complex-valued function of real variable t, then we can write

$$\int_a^b f(t)dt = \int_a^b u(t)dt + i\int_a^b v(t)dt, \quad \text{where } a \leq t \leq b.$$

Now we let $z(t) = x(t) + iy(t)$ where the curve determined by $z(t)$ is piecewise differentiable and we set $z'(t) = x'(t) + iy'(t)$. If x and y are continuous on $[a,b]$ and are continuously differentiable on some partition of $[a,b]$, we can define a line integral by

$$\int_C f(z)dz = \int_a^b f(z(t))z'(t)dt$$

where C is a piecewise smooth curve parametrised by $z(t)$.

If two curves C_1 and C_2 are smoothly equivalent then it turns out that

$$\int_{C_1} f = \int_{C_2} f.$$

If we define $-C$ by $z(b+a-t)$ (i.e. C traversed in the opposite direction) we find that

$$\int_{-C} f = \int_C f.$$

If f is the complex derivative of a holomorphic function F say, where F is holomorphic on C then

$$\int_C f(z)dz = F(z(b)) - F(z(a)).$$

If f is entire, then f is everywhere the derivative of a holomorphic function, namely there exists an entire F such that $F'(z) = f(z)$ for all z (proof of this is not given). We will use this and previous results to prove the first incarnation of the closed curve theorem.

B.4.2.2 Closed Curve Theorem. A curve is closed if its initial and end points coincide. If a curve C is given by $z(t)$ and $a \leq t \leq b$ then if $z(a) = z(b)$ C is closed. We will now show that the integral along a closed curve is zero.

Theorem 4.23

If f is entire and C is a smooth closed curve then
$$\int_C f(z)dz = 0.$$

Proof [Bak and Newman 82, 4.16, p 51]

Since f is entire there exists an $f(z) = F'(z)$ where F is an entire function. Hence
$$\int_C f(z)dz = \int_C F'(z)dz = F(z(b)) - F(z(a)).$$
Since C is closed, $z(a) = z(b)$ and $F(z(b)) = F(z(a))$ thus
$$\int_C f(z)dz = 0.$$

This can be related to Stokes' theorem by writing $\int_C f(z)dz$ as
$$\int_C (udx - vdy) = \oint \left(-\frac{dv}{dx} - \frac{du}{dy}\right) = 0$$
by the Cauchy–Riemann equations.

B.4.2.3 Cauchy Integral Formula. If f is entire and a is some given complex number and C is given by
$$C : Re^{i\theta}, \qquad 0 \leq \theta \leq 2\pi \qquad R > |a|$$
then
$$f(a) = \frac{1}{2\pi i} \int_C \frac{f(z)}{z-a} dz.$$
The proof of this theorem is not given here (see [Bak and Newman 82, 5.5, p 55]).

B.4.3 Liouville's Theorem and Fundamental Theorem of Algebra

B.4.3.1 Liouville's Theorem. An application of the Cauchy integral formula is Liouville's theorem. This says that if f is holomorphic and bounded in $\mathbf{C}$, then f is a constant. To prove this we simply calculate $f(a) - f(b)$ using the formula where a and b are complex numbers and C is a circle centred at zero with $R > (|a||b|)$. We note that we can take R as large as we want so $f(a) - f(b)$ approaches zero as R tends to infinity, i.e. $f(a) - f(b)$ which implies that f is a constant.

B.4.3.2 Fundamental Theorem of Algebra. Every non-constant polynomial with complex coefficients has a zero in **C**.

Proof [Bak and Newman 82, 5.12, p 59]

Let $P(z)$ be a polynomial. If $P(z) \neq 0$ for all z then we can write $f(z) = 1/P(z)$ which is an entire function. If $P(z)$ is not constant, then $P \to \infty$ as $z \to \infty$ so f is bounded. If f is bounded, then f is constant by Liouville's theorem. Hence P must be constant which contradicts the original assumption.

B.4.4 Properties of Holomorphic Functions

If f is holomorphic in an open disc $D(\alpha : r)$ and $\alpha \in D(\alpha : r)$ it can be shown that there exists functions F and G which are holomorphic in D such that

$$F'(z) = f(z) \quad \text{and} \quad G'(z) = \frac{f(z) - f(a)}{z - a}.$$

If C is a closed curve in D then

$$\int_C f(z)dz = \int_C \frac{f(z) - f(a)}{z - a} dz$$

and

$$\int_C \frac{f(z) - f(a)}{z - a} dz = \int_C G'(z)dz = G(z(b)) - G(z(a)).$$

Using this and the result that $\int_C \frac{dz}{z - a} = 2\pi i$ (proof [Bak and Newman 82, 5.4, p 55]) gives us Cauchy's integral formula for functions holomorphic in the disc D

$$f(a) = \frac{1}{2\pi i} \int_{C_P} \frac{f(z)}{z - a} dz, \quad 0 < p < r$$

where $C_P : \alpha + pe^{i\theta}$, $0 < \theta < 2\pi$.

Cauchy's theorem leads to an integral representation of derivatives of f, namely

$$f^{(n)}(a) = \frac{n!}{2\pi i} \int_{C_P} \frac{f(z)}{(z - a)^{n+1}} dz.$$

This in turn leads to a power series representation for holomorphic functions in $D(\alpha : r)$

$$f(z) = \sum_{n=0}^{\infty} C_n (z - a)^n.$$

We can choose $\alpha \in D(\alpha : r)$ and set $p > 0$ such that $|a - \alpha| < p < r$. If $|z - \alpha| < |a - \alpha|$ then

$$f(z) = \frac{1}{2\pi i} \int_{C_P} \frac{f(w)}{w - z} dw,$$

and using the result that

$$\frac{1}{w - \alpha} + \frac{z - \alpha}{(w - \alpha)^2} + \frac{(z - \alpha)^2}{(w - \alpha)^3} + \dots$$

converges to $1/(w - z)$ throughout C_P we get the following convergent series expansion for f:

$$\begin{aligned} f(z) &= \frac{1}{2\pi i} \int_{C_P} f(w) \left[\frac{1}{w - \alpha} + \frac{z - \alpha}{(w - \alpha)^2} + \frac{(z - \alpha)^2}{(w - \alpha)^3} + \dots \right] dw, \\ &= C_0(p) + C_1(p)(z - \alpha) + C_2(z - \alpha)^2 + \dots \end{aligned}$$

where

$$C_n(p) = \frac{1}{2\pi i} \int_{C_P} \frac{f(z)}{(z - \alpha)^{n+1}} dz.$$

As f is infinitely differentiable at α and $C_n(p) = \dfrac{f^{(n)}(\alpha)}{n!}$ for each p, $0 < p < r$ then

$$C_n = \frac{f^{(n)}(\alpha)}{n!} = \frac{1}{2\pi i} \int_{C_P} \frac{f(z)}{(z - \alpha)^{n+1}} dz$$

for all $z \in D(\alpha : r)$.

B.4.4.1 Mean Value Property. A consequence of Cauchy's theorem is the mean value property of holomorphic functions. This is where $f(\alpha)$ is equal to the mean value of f taken around the boundary of a disc centred at α, that is to say

$$f(\alpha) = \frac{1}{2\pi} \int_0^{2\pi} f(\alpha + re^{i\theta}) d\theta.$$

To prove this we simply reformulate Cauchy's integral formula and set $a = \alpha$.

B.4.4.2 Morera's Theorem. The converse of Cauchy's theorem is Morera's theorem which states that if f is continuous in some open subset of $\mathbf{C}$, D say and if

$$\int f = 0$$

then f is holomorphic in D. The proof of this theorem is simply to observe that we can define an indefinite integral F of f and then note that f is the derivative of the holomorphic function f.

B. Project Example 2: Algebraic Curves

B.4.5 General Cauchy Closed Curve Theorem

A function f can be holomorphic on a closed curve C and yet $\int_C f \neq 0$, for example

$$\int_{|z|=1} \frac{1}{z} dz = 2\pi i.$$

For the general case of Cauchy's closed curve theorem we want to find where the theorem is valid.

B.4.5.1 Simply Connected Domains. The function $f(z) = 1/z$ is the punctured plane, i.e. there is a 'hole' at $z = 0$. We want to define a region with no "holes". For this we use the definition of a simply connected domain. A domain is *simply connected* if every closed curve in the region can be contracted to a point of **C**. Clearly a rectangle is simply connected since every closed curve in the rectangle can be contracted to a point by means of affine transformations. Simply connected domains are best illustrated by simple examples (see below).

Example 4.52

The annulus $A = \{z : 1 < |z| < 3\}$ is not simply connected (see Figure B.7)

Example 4.53

The infinite strip $S = \{z : -1 < Im(z) < 1\}$ is simply connected (see Figure B.8)

We can now define the general case of Cauchy's theorem.

B.4.5.2 General Closed Curve Theorem. If f is holomorphic and lies in a simply connected domain D and C is a continuous smooth curve in D then

$$\int_C f = 0.$$

B.4.6 Isolated Singularities and Removable Singularities

A complex function f has an isolated singularity at z_0 if f is holomorphic in a deleted neighbourhood of z_0 but is not necessarily holomorphic at z_0, i.e. $f(z) = 1/(z-4)$ has a singularity at $z_0 = 4$.

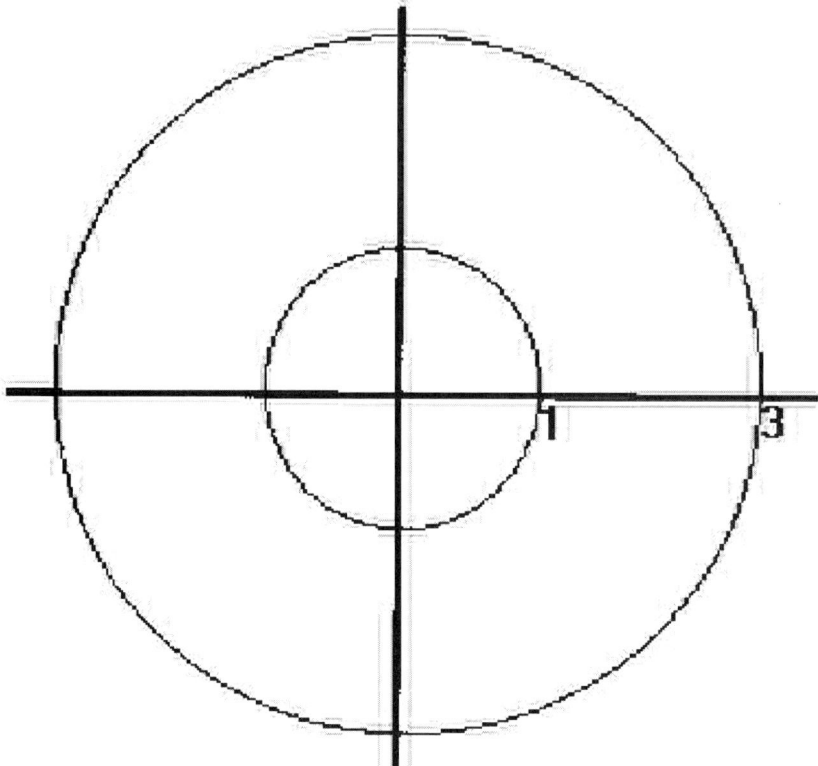

Figure B.7

Now suppose that f has a singularity at z_0. If there exists a holomorphic function g holomorphic at z_0 too so that $f(z) = g(z)$ for all z in some deleted neighbourhood of z_0 then we can say that f has a *removable singularity* at z_0, for example

$$f(z) = \begin{cases} \sin z & z \neq 2 \\ 0 & z = 2 \end{cases}$$

has a removable singularity at $z_0 = 2$.

If f can be written in the form $\dfrac{A(z)}{B(z)}$ where A and B are holomorphic and $A(z_0) \neq 0$ and $B(z_0) = 0$ then we say that f has a *pole* at z_0. If f does not have a pole or removable singularity at z_0 then it is said to have an *essential singularity* at z_0.

B. Project Example 2: Algebraic Curves

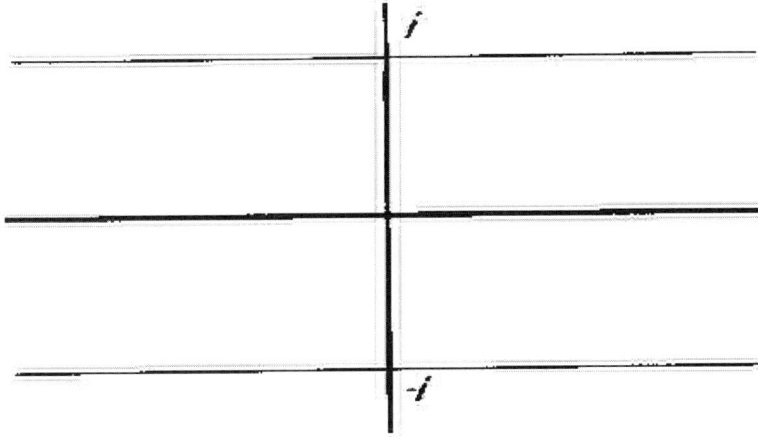

Figure B.8

B.4.7 Laurent Expansions

In section B.4.4 we saw that functions holomorphic in a disc could be represented by a power series. A *Laurent expansion* is a power series which involves negative powers of z also and which converges to a holomorphic function in an annulus $R_1 < |z - z_0| < R_2$, that is

$$f(z) = \sum_{-\infty}^{\infty} a_k (z - z_0)^k.$$

Conversely, if f is holomorphic in the annulus then it has a convergent power series as above with the coefficients a_k given by

$$a_k = \frac{1}{2\pi i} \int_C \frac{f(z)}{(z - z_0)^{k+1}} dz$$

where $C = C(z_0 : R)$ and R is chosen so that $R_1 < |z| < R < R_2$.

B.4.8 Residue Theorem

If γ is a circle surrounding z_0 and $f(z) = \sum_{-\infty}^{\infty} C_k (z - z_0)^k$ in a deleted neighbourhood of z_0 that contains γ then $\int_\gamma f = 2\pi i C_{-1}$. Thus the coefficient C_{-1} has a special importance and it is called the *residue* of f at z_0. If f has a pole of order one at z_0 and can be written $A(z_0)/B(z_0)$ then

$$C_{-1} = \text{Res}(f : z_0) = \lim_{z \to z_0} (z - z_0) f(z) = \frac{A(z_0)}{B'(z_0)}.$$

Example 4.81

$\frac{1}{(z^4-1)}$ has a pole at $z_0 = i$ and hence $\text{Res}(\frac{1}{(z^4-1)}; i) = \frac{1}{4i^3} = \frac{i}{4}.$

We can now generalise the Cauchy closed curve theorem further to include functions with singularities. To do this we must first consider the winding number of a closed curve. The winding number is given by

$$n(\gamma, a) = \frac{1}{2\pi i} \int_\gamma \frac{dz}{z - a}$$

where γ is a closed curve and a does not lie on γ. It can be shown that for any closed curve the winding number is an integer. (See the examples in Figure B.9). If we fix γ and vary a then the winding number is always a continuous

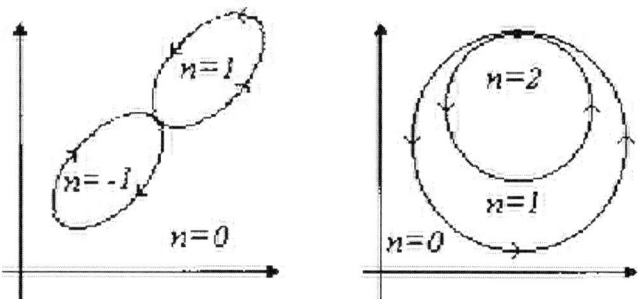

Figure B.9

function. If γ is a semi-circle traversed anti-clockwise then

$$n(\gamma, a) = \begin{cases} 1 & a \text{ inside semi-circle} \\ 0 & a \text{ outside semi-circle}. \end{cases}$$

B.4.8.1 Cauchy's Residue Theorem. If f is a holomorphic function in a simply connected domain except at isolated singularities $z_1, z_2, \ldots, z_m$ and if γ is a closed curve not intersecting any singularities then

$$\int_\gamma f = 2\pi i \sum_{k=1}^{m} n(\gamma, z_k) \text{Res}(f, z_k).$$

B. Project Example 2: Algebraic Curves

For proof see [Bak and Newman 82, 10.5, p 110]. We define f as *meromorphic* in a domain if f is holomorphic there except at isolated poles. A useful application of the residue theorem is the evaluation of definite integrals using contour integral techniques. Suppose we have an integral of the form

$$\int_{-\infty}^{\infty} \frac{P(x)}{Q(x)} dx$$

where P and Q are polynomials. This will make sense if $Q(x) \neq 0$ and if $deg Q - deg P \geq 2$, we then take

$$\int_{-\infty}^{\infty} \frac{P(x)}{Q(x)} dx = \lim_{R \to \infty} \int_{-R}^{R} \frac{P(x)}{Q(x)} dx$$

and we want to estimate this for large R. Let C_R be the closed curve from $-R$ to R of the real line. By the residue theorem we get

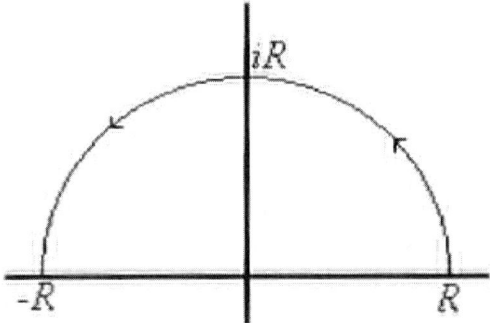

Figure B.10

$$\int_{C_R} \frac{P(z)}{Q(z)} dz = 2\pi i \sum_k \text{Res}(\frac{P}{Q}; z_k)$$

where z_k are poles of Q contained in C_R. It can be shown that when $R \to \infty$ we get

$$\int_{-\infty}^{\infty} \frac{P(x)}{Q(x)} dx = 2\pi i \sum_k \text{Res}(\frac{P}{Q}; z_k).$$

Example 4.83

$$\int_{-\infty}^{\infty} \frac{dx}{x^4+1} = 2\pi i \sum_k \text{Res}(\frac{1}{z^4+1}; z_k)$$

and we have poles at $z_1 = e^{\pi i/4}$ and $z_2 = e^{3\pi i/4}$. Hence

$$\text{Res}\left(\frac{1}{z^4+1}; e^{\pi i/4}\right) = \frac{1}{4z_1^3} = -\frac{1}{8}(\sqrt{2}+i\sqrt{2})$$

and

$$\text{Res}\left(\frac{1}{z^4+1}; e^{3\pi i/4}\right) = \frac{1}{4z_2^3} = -\frac{1}{8}(\sqrt{2}-i\sqrt{2})$$

thus

$$\int_{-\infty}^{\infty} \frac{dx}{x^4+1} = \frac{\pi\sqrt{2}}{2}.$$

B.4.9 Conformal Mapping

If f is holomorphic in some domain D which contains the point z_0 and if $f'(z_0) \neq 0$ then f is *conformal* at z_0. This can be interpreted as the angle between any two curves crossing at z_0 being preserved under f. For example $f(z) = e^z$ has a non-zero derivative at all points and is hence conformal. Under f the lines $x =$ constant and $y =$ constant must remain orthogonal. Under f the lines $x =$ constant map to circles centred at the origin and $y =$ constant maps to rays away from the origin (see Figure B.11). If f is a 1–1 holomorphic function

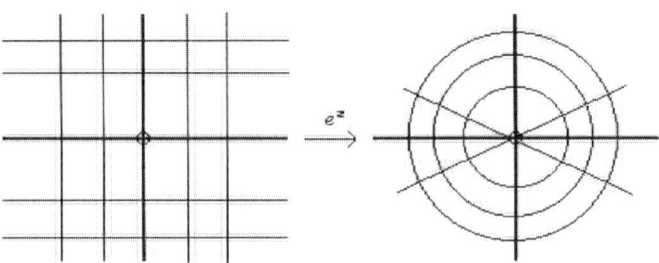

Figure B.11

in some domain D then f^{-1} exists in $f(D)$; also f and f^{-1} are conformal in

D and $f(D)$. A 1–1 holomorphic mapping is called a *conformal mapping*. This idea of a holomorphic map under some complex function will be used in the next section.

B.5 Topology and Riemann Surfaces

In the first chapter we looked at sketches of algebraic curves in $\mathbf{R}^2$. However we cannot sketch curves in $\mathbf{C}^2$ which has four real dimensions, but curves in $\mathbf{C}^2$ do have a topology and we can get accurate topological pictures.

B.5.1 Topology of Complex Algebraic Curves

Consider the standard form for the non-singular cubic which we investigated in B.3.2.1, namely $y^2 = g(x)$ where $g(x)$ is a cubic polynomial. Under a suitable projective transformation $g(x)$ can be written $g(x) = x(x-1)(x-\lambda)$. If we add a point at infinity then a non-singular cubic C is given by

$$C = \{y^2 = x(x-1)(x-\lambda)\} \cup \{\infty\}$$

where C is a subset of $\mathbf{CP}_2$ and $\lambda \in \mathbf{C} - \{0\}$. We want to investigate the topology of this object so consider the mapping

$$\pi : C \to \mathbf{CP}_2 \text{ defined by } \pi[x,y,z] \to [x,z] \text{ and } \infty \to [1,0].$$

This type of mapping is called a *branched covering* (i.e a double covering of C). The points which coincide in C are called *branch points*. In affine coordinates this is defined by $(x,y) \to x$ which is a 2–1 map corresponding to $y = \pm\sqrt{x(x-1)(x-\lambda)}$. Under stereographic projection $\mathbf{CP}_1$ is homeomorphic to the Riemann sphere. If we consider y as a function of x on $\mathbf{CP}_1$ then this has two values outside $[0,1,\lambda,\infty]$. Now we can cut $\mathbf{CP}_1$ along the two paths from 0 to 1 and from λ to ∞. This causes the double cover of C to fall apart into two pieces with y being single-valued on each sheet. We now open up the cuts. So a non-singular cubic in $\mathbf{CP}_2$ is topologically a torus. More generally it can be shown that any non-singular algebraic curve in $\mathbf{CP}_2$ is homeomorphic to a sphere with g handles where g is the *genus* of the curve and related to the degree d of the curve by the degree-genus formula

$$g = \frac{1}{2}(d-1)(d-2).$$

What is the topology of singular projective curves? We will look at what

$$C \cong$$

Figure B.12

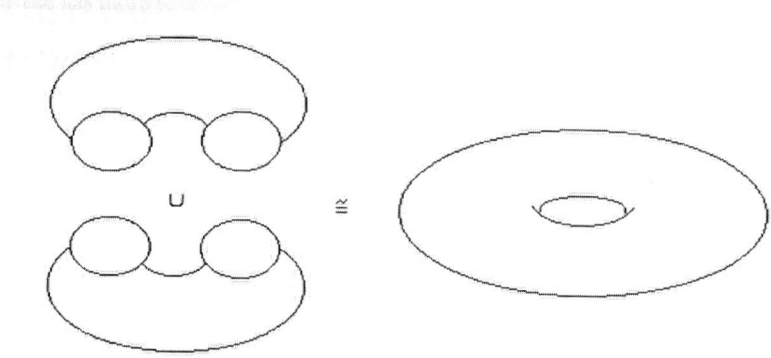

Figure B.13

happens at the end of this section. Thus a non-singular projective curve is topologically a surface in $\mathbf{CP}_2$. However this type of surface has more structure and we can do complex analysis on it. This type of surface is called a *Riemann surface*.

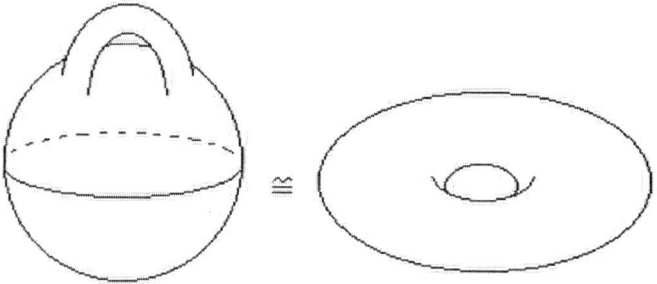

Figure B.14

B.5.2 Riemann Surfaces

To define a Riemann surface we first need to show what we mean by a topological surface. A topological surface is a space S which is locally homeomorphic to $\mathbf{C}$ or $\mathbf{R}^2$. A homeomorphism $\phi : U \to V$ between an open subset U of a surface S and an open subset of V of $\mathbf{C}$ is called a *chart* (see Figure B.15). A collection

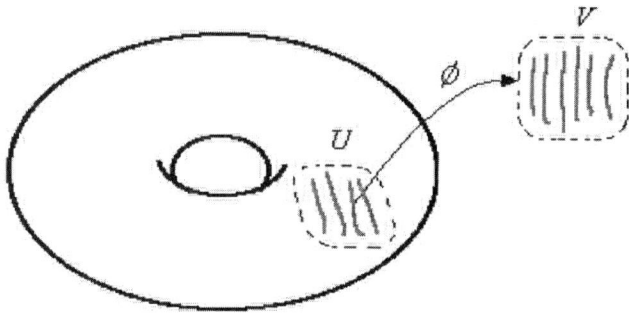

Figure B.15

of all charts on S is called an *atlas* Φ where $\Phi = \{\phi_\alpha : U_\alpha \to V_\alpha : \alpha \in A\}$ and hence for all $\alpha \in A$ S is the union of all U_α.

If we have the intersection of two open subsets in S then

$$\phi_\alpha : U_\alpha \to V_\alpha \text{ and } \phi_\beta : U_\beta \to V_\beta$$

are two charts and

$$\phi_\alpha(U_\alpha \cap U_\beta) \to V_\alpha \text{ and } \phi_\beta(U_\alpha \cap U_\beta) \to V_\beta.$$

We can now construct the homeomorphism

$$\phi_{\alpha\beta} = \phi_\alpha o \phi_\beta^{-1} : V_\beta \to V_\alpha$$

which maps between two open subsets of $\mathbf{C}$. This is known as the *transition function* (see Figure B.16). A surface is a Riemann surface if the transition

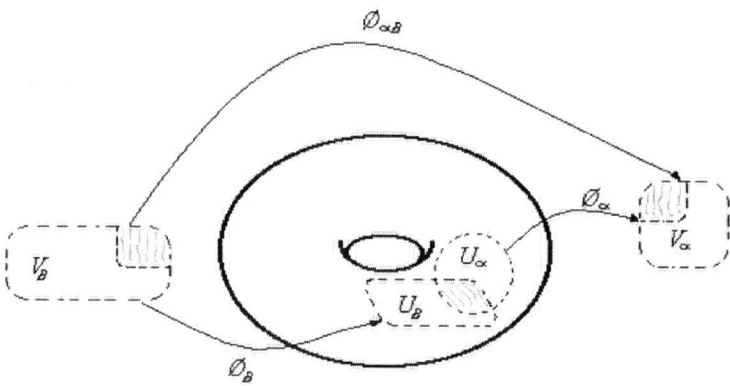

Figure B.16

functions are holomorphic (see section B.4.1) whenever they are defined. The atlas is then called a holomorphic atlas.

Example 5.21

The most obvious example of a Riemann surface is the Riemann sphere $\mathbf{CP}_1 = \mathbf{C} \cup \{\infty\}$. We let $U = \mathbf{CP}_1 - \{\infty\}$ and $V = \mathbf{CP}_1 - \{0\}$ and we define

$$\phi : U \to \mathbf{C} \text{ by } \phi[x,y] = x/y \text{ and } \psi[x,y] = y/x$$

(see Figure B.17). So ϕ and ψ form a holomorphic atlas on $\mathbf{CP}_1$ and hence

$$\phi o \psi^{-1} = \psi o \phi^{-1} : \mathbf{C} - \{0\} \to \mathbf{C} - \{0\}$$

B. Project Example 2: Algebraic Curves

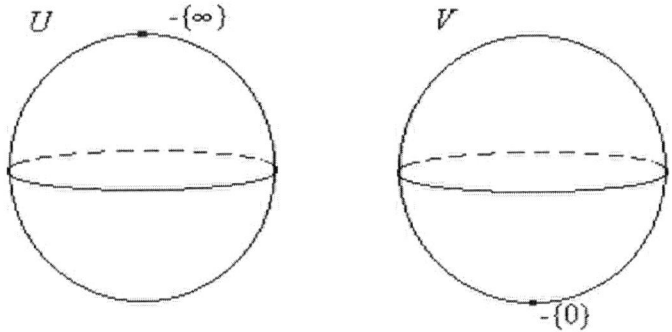

Figure B.17

are the transition functions defined by

$$z \to \frac{1}{z}$$

where $z \in \mathbf{C}$ are identified with $[z, 1] \in \mathbf{CP}_1$.

We can now return to algebraic curves and show that a projective curve $C - Sing(C)$ defined by $P(a, b, c)$ in $\mathbf{CP}_2$ forms a holomorphic atlas and is hence a Riemann surface. Suppose that $[a, b, c] \in C$ and also

$$P(a, b, c) = 0 \quad \text{and} \quad \frac{\partial P}{\partial y}(a, b, c) \neq 0.$$

By Euler's relation

$$a\frac{\partial P}{\partial x}(a, b, c) + b\frac{\partial P}{\partial y}(a, b, c) + c\frac{\partial P}{\partial z}(a, b, c) = 0$$

which implies that $a = c = 0$ and hence that $a = b = c = 0$ which is impossible by the definition of $C - Sing(C)$ as $[a, b, c]$ would be a singularity. So either $a \neq 0$ or $c \neq 0$. Assume that c is not equal to zero, then as P is homogeneous we can write

$$\frac{\partial P}{\partial y}(a/c, b/c, 1) = c^{-(d-1)}\frac{\partial P}{\partial y}(a, b, c) \neq 0$$

where d is the degree of the $P(x, y, z)$. The implicit function theorem applied to $P(x, y, 1)$ reveals that there are open neighbourhoods V and W of a/c and b/c in $\mathbf{C}$ and a holomorphic function $g : V \to W$ such that if $x \in V$ and $y \in W$ then

$$P(x, y, 1) = 0 \Leftrightarrow y = g(x).$$

If V and W are small then

$$U = \{[x,y,z] \in C : z \neq 0, x/z \in V, y/z \in W\} = \{[x,y,1] \in C\, x \in V, y \in W\}$$

is an open neighbourhood of $[a,b,c]$ in $C - Sing(C)$. The map $\phi U \to V$ defined by $\phi[x,y,z] = x/z$ has inverse $w \propto [w, g(w), 1]$. Similarly if $[a,b,c] \in C$ and we make the suppositions

$$\frac{\partial P}{\partial y}(a,b,c) \neq 0 \neq a \quad \text{or} \quad \frac{\partial P}{\partial x}(a,b,c) \neq 0 \quad \text{or} \quad \frac{\partial P}{\partial z}(a,b,c) \neq 0$$

then by similar methods we can find a homeomorphism $\phi U \to V$ from an open subset V of $\mathbf{C}$ where $\phi[x,y,z]$ has one of these forms

$$z/x, y/z, z/y, x/y, y/x.$$

The inverses have the following forms:

$$w \to [1, g(w), w], [g(w), w, 1], [g(w), 1, w], [w, 1, g(w)] \text{ or } [1, w, g(w)].$$

Now consider the homeomorphism $\phi_\alpha[x,y,z] = x/z$ which has inverse $[w, g(w), 1]$ and $\phi_\beta[x,y,z] = y/x$ which has inverse $[1, w, g(w)]$. Hence,

$$\phi_\alpha : U_\alpha \to V_\alpha \quad \text{and} \quad \phi_\beta : U_\beta \to V_\beta$$

where U_α and U_β are open neighbourhoods of $[a,b,c]$ intersecting in $C - Sing(C)$ and V_α and V_β are open subsets of V. We can construct transition functions

$$\phi_{\alpha\beta} = \phi_\alpha o \phi_\beta^{-1} : V_\alpha \to V_\beta \quad \text{and} \quad \phi_{\beta\alpha} = \phi_\beta o \phi_\alpha^{-1} : V_\beta \to V_\alpha.$$

See Figure B.18, hence $\phi_{\alpha\beta} = 1/g(w)$ and $\phi_{\beta\alpha} = g(w)/w$.

If we define $g : V \to \mathbf{C}$ to be holomorphic then the transition functions are holomorphic. Similarly if we adopt this method for our other forms of ϕ then we get transition functions to be one of the following:

$$w \to w, 1/w, g(w), 1/g(w), w/g(w), g(w)/w$$

where g is holomorphic and does not vanish where the transition function is defined. Consequently we get a holomorphic atlas on $C - Sing(C)$ and hence $C - Sing(C)$ is a Riemann surface.

B. Project Example 2: Algebraic Curves

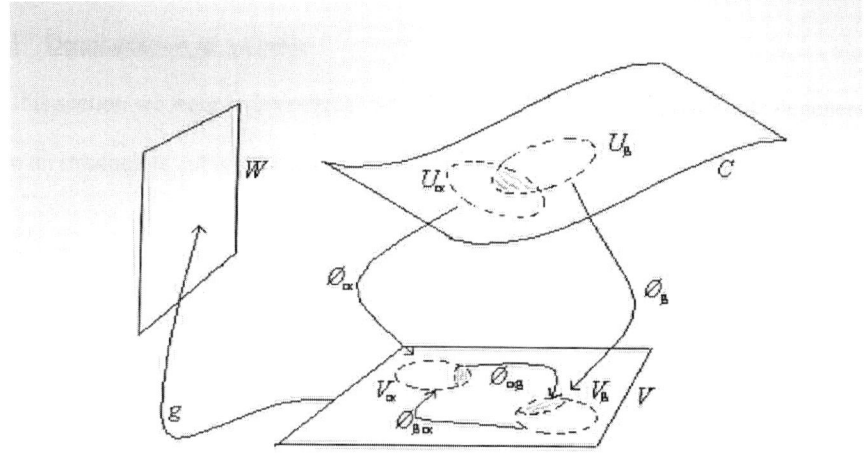

Figure B.18

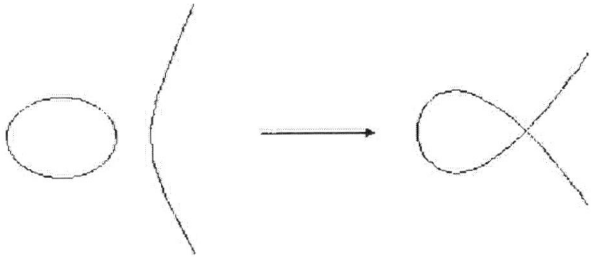

Figure B.19

B.5.3 Degeneration of a Cubic

In this section we want to investigate what happens as the non-singular cubic degenerates into an irreducible cubic with a node (see Figure B.19). Consider a curve C described by $C = \{y^2 = x^3 + x^2 - c\}$ where c is real and small. As $c \to 0$, C degenerates into a cubic with an ordinary double point. To investigate what happens we again consider the mapping $\pi : C \to \mathbf{CP}_1$ namely a branch covering. In affine coordinates this corresponds to the map $y = \pm\sqrt{x^2(x+1) - c}$ which has three branch points, a, b, c say, plus a point at infinity. We know that $\mathbf{CP}_1 = \mathbf{C} \cup \{\infty\}$ so we represent the branch points on $\mathbf{C}$ by Figure B.20. At $c = 0$ under π we get the map corresponding to $y = \pm\sqrt{(x+1)}$ which has two branch points at -1 and ∞. So as c tends to

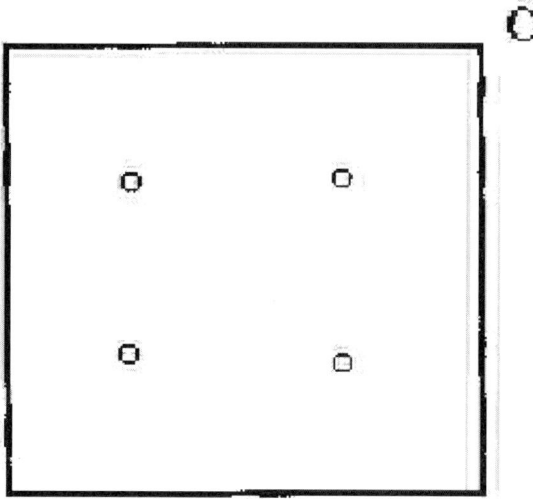

Figure B.20

zero, four branch points tend towards two branch points, see Figure B.21. It follows that when $c = 0$ we get the Riemann sphere identified with two points. From Section B.5.1, we know that a non-singular cubic is topologically a torus and we can represent the branch points a, b, c and ∞ on the edge of a torus. Topologically a singular cubic with a node can be represented by contracting the meridian a, b to a point. This is topologically equivalent to the Riemann sphere identified with two points, Figure B.22. As $c \to 0$ we get a *vanishing cycle* for a cubic, i.e. a circle around the meridian of a torus contracting to a point also represented in the next figure (Figure B.23) by the real cone. These pictures although topologically correct do not represent how the curve lies in $\mathbf{C}^2$. In the case of the cubic with a node in $\mathbf{R}^2$ the singular point is two lines passing through the origin. However the complex algebraic curve at the singular point consists of two planes intersecting transversely in $\mathbf{C}^2$ which is not possible to draw adequately.

We also notice that the genus of the curve changes from one to zero. In general the genus g of a curve of degree d with ordinary singular points of multiplicity m_i is

$$g = \frac{1}{2}(d-2)(d-1) - \sum \frac{1}{2} m_i(m_i - 1).$$

B. Project Example 2: Algebraic Curves

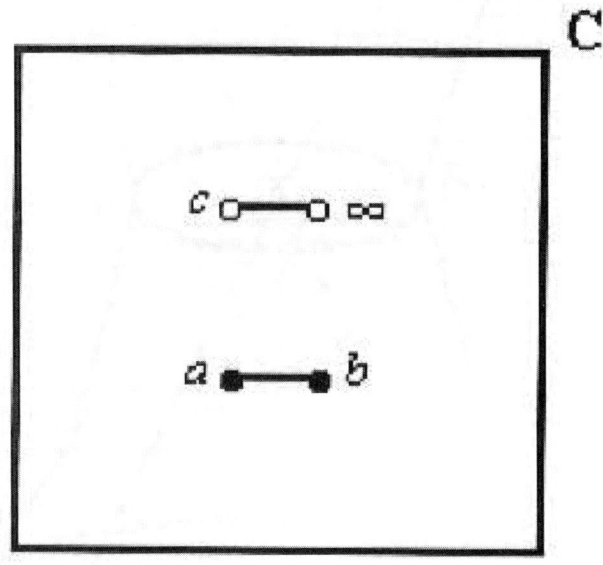

Figure B.21

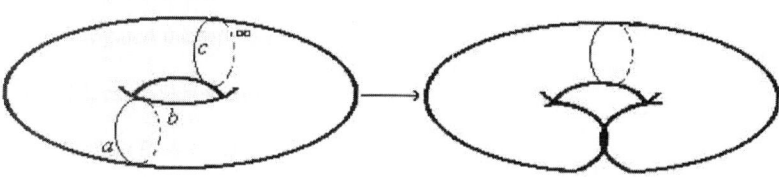

Figure B.22

B.5.4 Singularities and Riemann Surfaces

We have investigated the topology of singular curves but what happens at the singularities? It can be shown for any singular projective curve C in $\mathbf{CP}_2$ we can construct a surjective map $\pi : R \to C$ where R is a compact Riemann surface. The singular points of C are the image of a finite set of points in R

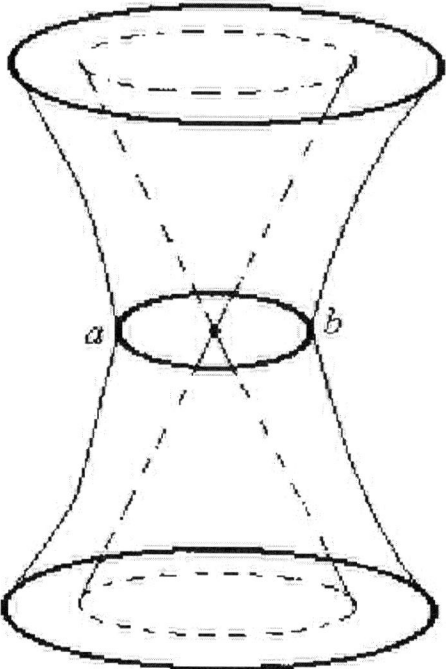

Figure B.23

under π, and on the complement of this set π defines a holomorphic bijection

$$\pi : R - \pi^{-1}(Sing(C)) \to C - Sing(C).$$

We call this a *resolution of singularities*.

B.6 Further Topics

In this final section we briefly look at some other ideas which lead on directly from what we have learnt in complex analysis. The discussion is informal and proofs are hence omitted.

B.6.1 The Weierstrass Function

We let $L = \{nw_1 + mw_2 : n, m \in \mathbf{Z}\}$ where L is a lattice in $\mathbf{C}$ which inherits a natural group structure of $\mathbf{C}$, see Figure B.24. There is a meromorphic function

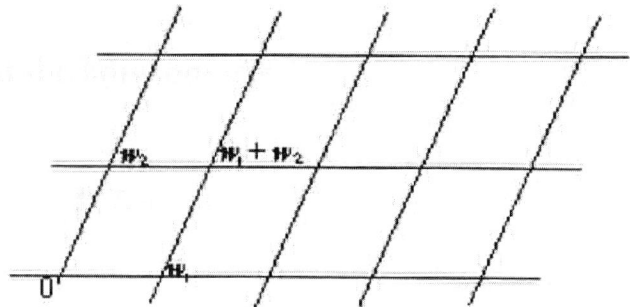

Figure B.24

in $\mathbf{C}$ defined by

$$\wp(z) = z^{-2} + \sum_{w \in L - \{0\}} [(z-w)^{-2} + w^{-2}]$$

with derivative

$$\wp'(z) = \sum_{w \in L} -2(z-w)^{-3}$$

where $\wp(z)$ is called the *Weierstrass function* associated with the lattice L and $w = nw_1 + mw_2$, i.e. we can choose any n and m to obtain w.

It can be shown that $\wp$ satisfies the identity

$$\wp'(z) = 4\wp(z)^2 - g_2\wp(z) - g_3$$

where

$$g_2 = g_2(L) = 60 \sum_{w \in L-\{0\}} w^{-4}, \text{ and } g_3(L) = 140 \sum_{w \in L-\{0\}} w^{-6}.$$

Using this identity we can define a non-singular projective curve C_L by

$$y^2 z - 4x^3 + g_2 x z^2 + g_3 z^3 = 0$$

with g_2 and g_3 as before.

We know that the lattice L has an additive group structure and is a subgroup of $\mathbf{C}$ so we can form a quotient group

$$\mathbf{C}/L = \{L + \alpha : L \in \mathbf{Z}\}.$$

We can investigate the topology of $\mathbf{C}/L$ by the map $\pi : \mathbf{C} \to \mathbf{C}/L$. The effect of this map is to "glue" appropriate etches of the parallelogram together. Hence $\mathbf{C}/L$ is topologically a torus.

It can also be shown that the homeomorphism $\phi : \mathbf{C}/L \to C_L$ defined by

$$\phi(L+z) = \begin{cases} [\wp(z), \wp'(z), 1] & z \notin L \\ [0,1,0] & z \in L \end{cases}$$

is holomorphic. So we can associate a lattice L in $\mathbf{C}$ to a non-singular cubic curve C_L in $\mathbf{CP}_2$ and conversely it can be shown that every non-singular cubic curve comes from some C_L.

If we are given C_L how do we get the lattice L? To do this we need to integrate along a smooth path in a Riemann surface.

B.6.2 Differential Forms on a Riemann Surface

To define integration on a curve we need to show what we mean by a smooth path. A piecewise smooth path in a Riemann surface S is a continuous map γ from a closed interval $[a,b]$ in $\mathbf{R}$ to S such that if $\phi : U \to V$ is a holomorphic chart on an open subset U of S and $[c,d] \subseteq \gamma^{-1}(U)$ then $\phi o \gamma : [c,d] \to V$ is a piecewise smooth path in the open subset V of $\mathbf{C}$. We have a closed path if $\gamma(a) = \gamma(b)$.

In order to integrate in S we need to define the concept of meromorphic and holomorphic differentials. A meromorphic function on a Riemann surface S is a function $f : S \to \mathbf{CP}_1$ which is holomorphic and is not identically ∞ on any connected component of S.

Let f and g be meromorphic functions on an open subset of a Riemann surface S. We define the symbol fdg as a meromorphic differential on S. If $\tilde{f}$ and $\tilde{g}$ are meromorphic functions on S then we say that $fdg = \tilde{f}d\tilde{g}$ if and only if for every holomorphic chart $\phi : U \to V$ on an open subset U of S we get

$$(fo\phi^{-1})(go\phi^{-1})' = (\tilde{f}o\phi^{-1})(\tilde{g}o\phi^{-1})'.$$

We say that the meromorphic differential fdg has a pole at $\phi(p)$. If fdg has no poles then it is defined as a holomorphic differential.

The integral of a holomorphic differential on S along a piecewise smooth path $\gamma : [a, b] \to S$ is given by

$$\int_\gamma f dg = \int_a^b f o\gamma(t)(g o\gamma)' dt$$

where t is a parameter independent of the choices of $f dg$.

If the Riemann surface is $\mathbf{C}$ then

$$\int_\gamma f dg = \int_\gamma f(z) g'(z) dz$$

from the normal form in complex analysis.

B.6.2.1 Abelian Integrals. An *Abelian integral* is an integral of the form

$$\int_\gamma f dg$$

where f and g are rational functions on $C - Sing(C)$ and γ is a piecewise smooth path in $C - Sing(C)$ not passing through the poles of the meromorphic differential $f dg$. We assume that C is not the line at infinity defined by $z = 0$ and we take the rational function g to be $[x, y, z] \propto x/z$ and we hence work in affine coordinates $[x, y, 1]$ and write dx for dg. Consequently, f becomes a rational function $R(x, y)$ and the integral is written

$$\int_\gamma f dg = \int_\gamma R(x, y) dx.$$

A special case of an Abelian integral is given by the integral over the curve C_L as defined in Section B.6.1 where C_L is associated with the lattice L.

We define γ_1 and γ_2 to be closed smooth paths in $\mathbf{C}/L$, see Figure B.25. Any closed path on $\mathbf{C}/L$ can be obtained by using combinations of γ_1 and γ_2 and we can traverse these paths as many times as we want. It can be shown that there is a meromorphic differential η such that

$$L = n \int_{\gamma_1} \eta + m \int_{\gamma_2} \eta = n w_1 + m w_2 \qquad \text{where } n, m \in \mathbf{Z}.$$

In Section B.6.1 we defined the homeomorphism $\phi : \mathbf{C}/L \to C_L$ and if we work in affine coordinates η is a rational function and it can therefore be shown that

$$L = \int_\gamma y^{-1} dx$$

where γ is a piecewise smooth curve in C_L. Hence given any non-singular projective cubic we get a lattice L in $\mathbf{C}$. As $\mathbf{C}$ has a natural additive group structure it makes sense that the cubic has a group structure as discussed in Section B.3.3.

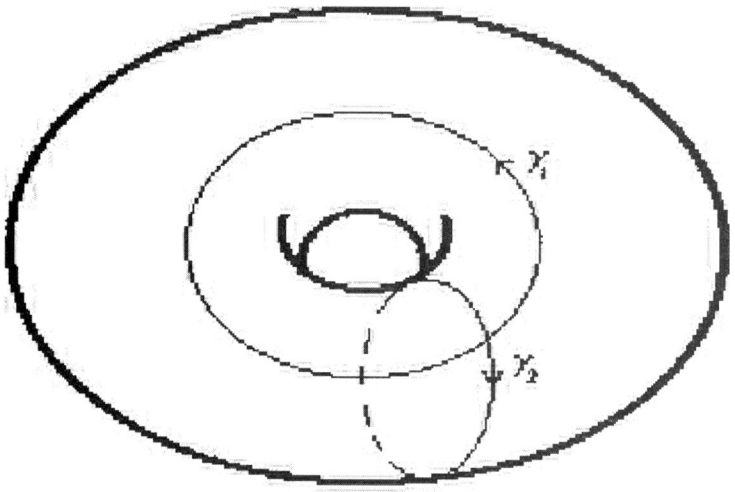

Figure B.25

B.6.3 Abel's Theorem

In Section B.6.1 we saw that a complex torus $\mathbf{C}/L$ is biholomorphic to the cubic C_L in $\mathbf{CP}_2$. Using this result and our investigation into differential forms leads to Abel's theorem for cubics which gives an addition formula for elliptic integrals.

Theorem 6.41

If $t, v, w \in \mathbf{C}$ then $t + v + w \in L$ if and only if there is a line in $\mathbf{CP}_2$ whose intersection with C_L consists of points $\phi(L+t), \phi(L+v)$ and $\phi(L+w)$ where ϕ is the homeomorphism defined in Section B.6.1.

Equivalently, if p, q, r lie in C_L then

$$L + \int_{[0,1,0]}^{p} y^{-1} dx + \int_{[0,1,0]}^{q} y^{-1} dx + \int_{[0,1,0]}^{r} y^{-1} dx = L + 0.$$

Proof[Kirwan 92, 6.23, p 155].

Bibliography

1. [Bak and Newman 82] Bak J. and Newman D.J., *Complex Analysis*, Undergraduate Texts in Mathematics Springer–Verlag, Berlin, 1982.

2. [Beardon 84] Beardon A.F., *A Primer on Riemann Surfaces* Lond. Math Soc Lecture Notes 78, Cambridge, 1984

3. [Brieskorn and Knorrer 86] Brieskorn E. and Knorrer H., *Plane Algebraic Curves*, Birkhauser-Verlag, Basel 1986.

4. [Kirwan 92] Kirwan F., *Complex Algebraic Curves*, Lond. Math Soc. Student Texts 23, Cambridge, 1992.

5. [Reid 88] Reid M., *Undergraduate Algebraic Geometry*, Lond. Math Soc. Student Texts 12, Cambridge, 1988.

6. [Siegal 71] Siegal C.L., *Topics in Complex Function Theory, Vols I and II*, Wiley Classics Library, New York, 1971.

7. [Walker 50] Walker R.J., *Algebraic Curves* Princeton University Press, Princeton, 1950.

Epilogue

This project report reads quite well, if not as well as that in Appendix A. The material is more accessible, indeed the chapters on complex variable could form part of the second year curriculum. In this instance this particular student could not take any modules on complex analysis other than elementary first year material on complex numbers. Some of the later ideas involving topology, Riemann surfaces and the Weierstrass function are quite advanced. The assessors for this project were considering the possibility of a good second class degree or perhaps going even higher. Then came the oral examination where two assessors asked the student questions on the written project report.

This was conducted in a friendly supportive way, but when it came to saying anything sensible, at the oral examination the student was completely tongue tied. There was no evidence that the student understood what had been written; even the more elementary questions were left unanswered. There are several reasons why this could happen. The nasty reason is that someone else really has written the project! If this can be proved, or the student confesses, a zero mark has to be awarded. For this project, this was not the case. Alternatively the student could be very nervous and, in the stressful situation of an oral, completely dry up. The student could be swept along during the year by an over enthusiastic supervisor, so much so that even the student thinks (s)he understands everything until the crunch comes and (s)he finds all questions at the oral baffling. A third reason could be that the student could be too laid back, and there is usually sufficient time between submitting a project report and the oral to forget everything, or at least put it beyond recall amongst all the other stuff needed for finals that are looming large. Whatever the reason, the student did him/herself no justice at all at the oral. The compromise mark was given in the mid 50s, corresponding to a lower second class degree. The external examiner was probably surprised the mark was this low until he read the reports from the assessors. It is very disappointing for both student and supervisor (first assessor) when this happens.

C
Project Example 3: Water Waves on a Sloping Beach

Preamble

In contrast to the first two appendices, this project example is very much applied mathematics. The mathematics is much more elementary, but there is the possibility of relating it to the real world. This project thus resembles the boomerang project outlined in Section 3.5 or the hurricane project of Section 3.6. Although the equations governing water waves are not in the course, their solution as presented here could be easily understood by second or even first year undergraduates. In order to get high marks in such a project therefore, it is necessary to either do some advanced mathematics, e.g. stochastic waves, non-linear waves, or to relate the simple mathematics to a real world problem and to understand the limitations of the model, how it might be extended, possible numerical work, etc. It will be seen that unfortunately this project does not really do this.

C.1 Abstract

In this project I have studied the basic Surface Water Wave equations to obtain a linear model for the Free Surface and Velocity Potential. From then on right running wave equations were used and from these the wave velocity and motion of the particle are calculated which could be used to illustrate a breaking wave in shallow water. After this wave rays were studied to be able to look at the direction of the waves on a sloping beach and show refraction of the wave crests as they approach the shore line.

C.2 Introduction

Waves on the ocean are generated by wind, which you can see as when there are high winds the sea is rough and on calm days the sea can be almost flat. Wind dislodges the surface of the water and gravity endeavours to put it back to its original state, nature tries to remain in equilibrium. These waves generated continue travelling on the open ocean, varying in size depending on conditions, they are dampened if opposing the direction of the wind or can grow and travel faster when going with the wind. They will obviously eventually meet such obstacles as land!

Something very interesting happens as the waves approach land, i.e. a beach, they refract and the wave crests turn parallel to the shoreline. You will also notice that the waves break on the beach.

I have read various books on this subject and found that *Ocean Engineering Wave Mechanics* by M.E. McCormick and *An Introduction to Water Waves* by G.D. Crapper to be the most helpful in my understanding of the subject.

To begin with we will calculate a wave equation for Linear Surface Waves, the free surface is modelled by η at some position x and time t, and $z = 0$ is assumed to be the still water level as illustrated in Figure C.1.

C.3 Surface Waves

The Free Surface is defined by

$$z = \eta(x, t). \tag{C.1}$$

The flow is assumed to be irrotational

$$\underline{\nabla} \times \underline{V} = 0. \tag{C.2}$$

Therefore the fluid potential $\underline{V}$ can be represented by a potential function ϕ

$$\underline{V} = \nabla \phi. \tag{C.3}$$

C. Project Example 3: Water Waves on a Sloping Beach

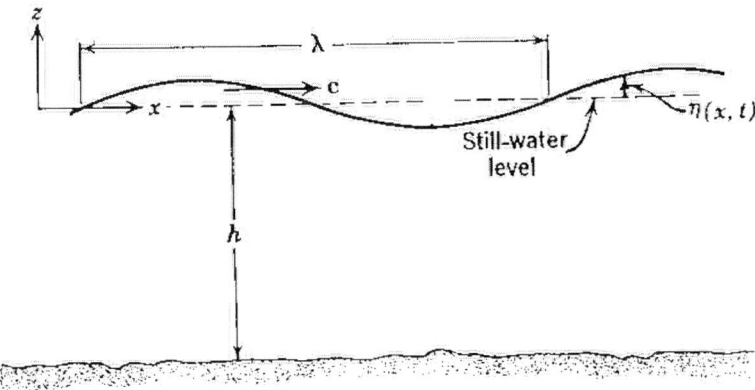

Figure C.1   Travelling surface wave. (Adapted from *Ocean Engineering Wave Mechanics* by M.E. McCormick, John Wiley and Sons Inc. Reprinted with permission.)

Continuity states that
$$\underline{\nabla}.\underline{V} = 0. \tag{C.4}$$
From Equation (C.3), (C.4) becomes
$$\nabla^2 \phi. \tag{C.5}$$
That is Laplaces Equation.

C.3.1 The Current

Suppose we impose a uni-directional current
$$\underline{u} = U\underline{i}. \tag{C.6}$$

This current produces additional velocity potential ϕ of the form
$$\phi_1 = Ux + \phi \tag{C.7}$$
$$\therefore \frac{\partial \phi_1}{\partial x} = U. \tag{C.8}$$

However here it will be assumed that there is no current.

C.3.2 The Boundary Conditions

Before we can solve Laplaces Equation there are some boundary conditions we need to impose.

Suppose that S is a surface in a fluid, moving with the fluid, and fluid already in S remains there.

Expressing S as the equation

$$S(x, z, t) = 0. \tag{C.9}$$

Also particles in S remain there with varying x, z and t which gives us

$$\frac{DS}{Dt} = 0 \tag{C.10}$$

where D denotes differentiation following the fluid.

Now with our free surface (C.1), S can be written as

$$S = \eta(x, t) - z = 0. \tag{C.11}$$

The Chain Rule states that

$$\frac{Ds}{Dt} = \frac{\partial s}{\partial t} + \frac{\partial s}{\partial x}\frac{dx}{dt} + \frac{\partial s}{\partial z}\frac{dz}{dt} \tag{C.12}$$

$\frac{dx}{dt}$ and $\frac{dz}{dt}$ are the components of the velocity $\underline{V}(u, , v, w)$ $(v = 0)$.

Therefore Equations (C.11) and (C.12) gives

$$\frac{Ds}{Dt} = \frac{\partial s}{\partial t} + \underline{V}.\nabla s = 0 \tag{C.13}$$

$$\Rightarrow \frac{\partial \eta}{\partial t} + u\frac{\partial \eta}{\partial x} - w = 0 \tag{C.14}$$

with

$$u = \frac{\partial \phi}{\partial x} \qquad w = \frac{\partial \phi}{\partial z}. \tag{C.15}$$

With no current and the second set of Equations (C.15), (C.14) gives

$$\frac{\partial \eta}{\partial t} = \frac{\partial \phi}{\partial z} \quad \text{at } z = \eta. \tag{C.16}$$

This is called the Kinematics Surface Condition

On the sea bed

$$z = -h(x). \tag{C.17}$$

Therefore Equation (C.13) gives

$$u\frac{\partial h}{\partial x} + w = 0 \quad \text{at } z = -h \tag{C.18}$$

C. Project Example 3: Water Waves on a Sloping Beach

and once again assuming no current, on the sea bed, (C.18) gives

$$\frac{\partial \phi}{\partial z} = 0 \quad \text{at } z = -h. \tag{C.19}$$

This is the sea bed boundary condition.

Euler's Equation of motion for inviscid flow is

$$\frac{\partial \underline{V}}{\partial t} + (\underline{V}.\nabla)\underline{V} = \frac{-1}{\rho}\nabla p. \tag{C.20}$$

From continuity, (C.4), the second term on the L.H.S of (C.20) is zero. Hydrostatic balance is

$$p = \rho g(\eta - z) \tag{C.21}$$

and recall that

$$\nabla p = (\frac{\partial p}{\partial x}, \frac{\partial p}{\partial y}, \frac{\partial p}{\partial z}). \tag{C.22}$$

Therefore hydrostatic balance implies

$$\frac{-1}{\rho}\frac{\partial p}{\partial x} = -g\frac{\partial \eta}{\partial x} \tag{C.23}$$

and Euler's Equation of motion becomes

$$\frac{\partial \underline{V}}{\partial t} = -g\frac{\partial \eta}{\partial x}. \tag{C.24}$$

Representing $\underline{V}$ as in (C.3), (C.24) can be written as

$$\frac{\partial}{\partial x}(\frac{\partial \phi}{\partial t}) = -g\frac{\partial \eta}{\partial x} \tag{C.25}$$

$$\Rightarrow \int \frac{\partial}{\partial x}\frac{\partial \phi}{\partial t}dx = \int -g\frac{\partial \eta}{\partial x} \tag{C.26}$$

$$\frac{\partial \phi}{\partial t} = -g\eta + \text{const.} \tag{C.27}$$

Taking the constant of integration to be zero, we obtain

$$\eta = -\frac{1}{g}\frac{\partial \phi}{\partial t} \quad \text{at } z = \eta. \tag{C.28}$$

This is called the Dynamic Surface Condition.

To avoid difficulties when solving Laplaces Equation, with the imposed boundary conditions, at $z = \eta$, it is assumed that the mean surface level is at $z = 0$.

C.3.3 A Separable Solution of Laplace's Equation

It is assumed that ϕ has a separable solution of the form

$$\phi = X(x)Z(z). \tag{C.29}$$

Applying Laplaces Equation to this gives us

$$\frac{1}{X}\frac{d^2X}{dx^2} = \frac{-1}{Z}\frac{d^2Z}{dz^2} = -k^2. \tag{C.30}$$

We can now solve for Z

$$\frac{d^2Z}{dz^2} = -k^2 \tag{C.31}$$

of which a general solution can be found to be

$$Z = C\sinh(kz) + D\cosh(kz). \tag{C.32}$$

Using the boundary condition on the sea bed, (C.19), we have

$$\frac{dZ}{dz} = 0 \quad \text{at } z = -h \tag{C.33}$$

where C and D are constants. Using condition (C.33) we can find that

$$C = D\frac{\sinh(kh)}{\cosh(kh)}. \tag{C.34}$$

Therefore we now have $Z(z)$ as

$$Z(z) = D\cosh(kz + kh) \tag{C.35}$$

Similarly for X

$$\frac{d^2X}{dx^2} + k^2X = 0. \tag{C.36}$$

We get the general solution

$$X(x) = A\sin(kx) + B\cos(kx) \tag{C.37}$$

where A and B are constants.

Now using the two surface conditions (C.16) and (C.28) we can obtain

$$\frac{\partial\phi}{\partial z} + \frac{1}{g}\frac{\partial^2\phi}{\partial t^2} = 0 \quad \text{at } z = 0. \tag{C.38}$$

From this we can find an expression for $T(t)$

$$TDk\sinh(kh) + \frac{1}{g}\frac{d^2T}{dt^2}D\cosh(kh) = 0 \tag{C.39}$$

C. Project Example 3: Water Waves on a Sloping Beach

Making

$$\omega = \sqrt{gk\tanh(kh)} \tag{C.40}$$

$$\frac{d^2T}{dt^2} + \omega^2 T = 0 \tag{C.41}$$

with the general solution being

$$T(t) == E\sin(\omega t) + F\cos(\omega t). \tag{C.42}$$

E and F are constants Letting B and F equal zero and putting together our X, (C.35), Z, (C.31), and T, (C.42) gives us the velocity potential

$$\phi = A\sin(\omega t)\sin(kx)\cosh(kz+kh). \tag{C.43}$$

The constants A, D and E are written as the one new constant A.

Now that we have a solution for ϕ we can use the Dynamic Surface condition, (C.28), to obtain an equation for the free Surface, η.

$$\eta = -\frac{A\mu\cosh(kh)\sin(kx)\cos(\omega t)}{g}. \tag{C.44}$$

Letting

$$a = -\frac{A\omega\cosh(kh)}{g}. \tag{C.45}$$

Using this in Equation (C.44) we have

$$\eta = a\sin(kx)\cos(\omega t). \tag{C.46}$$

Other solutions of this but with different combinations of cosine and sine are obtained when we allow, say, B and E to equal zero, (or A and E, or B and F)

$$\eta = a\sin(kx)\sin(\omega t) \tag{C.47}$$

$$\eta = a\cos(kx)\cos(\omega t) \tag{C.48}$$

$$\eta = a\cos(kx)\sin(\omega t). \tag{C.49}$$

Because the solutions we have obtained are for linear waves we can use the property of Superposition to add together two of our μ's to make left and right running waves.

$$\text{Left running wave} \quad \eta = a\sin(kx+\omega t) \tag{C.50}$$

$$\text{Right running wave} \quad \eta = a\cos(kx-\omega t). \tag{C.51}$$

We then have corresponding Velocity Potentials

$$\phi = A\cosh(kx+kh)\cos(kx+\omega t) \tag{C.52}$$

$$\phi = A\cosh(kz+kh)\sin(kx-\omega t). \tag{C.53}$$

We shall only be looking at right running waves.

C.4 [No Title]

C.4.1 The Velocity of the Waves

If the coordinate system is allowed to move with the argument of the cosine in Equation (C.51), then
$$kx - \omega t = \text{const.} \tag{C.54}$$
So we can see that the differentiation of this is equal to zero which leads us to the wave velocity, c
$$c = \frac{dx}{dt} = \frac{\omega}{k} \tag{C.55}$$
and using our value of ω from (C.40)
$$c = \sqrt{\frac{g \tanh(kh)}{k}}. \tag{C.56}$$

C.4.2 The Group Velocity of the Waves

This is the velocity of a group of waves with the same amplitude, a, but slightly varying wave number, k, and frequency. Take two waves with waves numbers k and δk, and frequencies with ω and $\delta \omega$. Then their combined velocities would be
$$c_g = \frac{\delta \omega}{\delta k} \tag{C.57}$$
and as δk and $\delta \omega$ tend to zero,
$$c_g = \frac{d\omega}{dk}. \tag{C.58}$$
Then differentiating ω from (C.40) we have
$$c_g = \frac{gkh\,\text{sech}^2(kh)}{2\omega}. \tag{C.59}$$

C.4.3 The Motion of the Particles

Remembering that the velocity components of $\underline{V}$ are
$$u = \frac{\partial \phi}{\partial x} \qquad v = 0 \qquad w = \frac{\partial \phi}{\partial z}. \tag{C.60}$$

Now that we have our velocity potential we obtain solutions to these

$$u = \frac{agk \cosh(kz + kh) \cos(kx - \omega t)}{\omega \cosh(kh)} \qquad (C.61)$$

$$w = \frac{agk \sinh(kz + kh) \sin(kx - \omega t)}{\omega \cosh(kh)}. \qquad (C.62)$$

The particles can take on new coordinates, (α, β), with the origin at $z = 0$ and some point $x = x_0$. So now u and w vary with time and we have

$$u = \frac{d\alpha}{dt} \qquad w = \frac{d\beta}{dt}. \qquad (C.63)$$

We can integrate u and w with respect to t to find expressions α and β

$$\alpha = \frac{agk \sin(kx_0 - \omega t)}{-\omega^2} \qquad (C.64)$$

$$\beta = \frac{agk \tanh(kh) \cos(kx_0 - \omega t)}{-\omega^2}. \qquad (C.65)$$

Equations (C.64) and (C.65) can be rearranged, squared and added together to give

$$\frac{\alpha^2}{\left(\frac{kag}{\omega^2}\right)^2} + \frac{\beta^2}{\left(\frac{kag \tanh(kh)}{\omega^2}\right)^2} = 1. \qquad (C.66)$$

Thus from this we can see that the motion of the particles is an Ellipse which is affected in shape by the depth of the water, h In very deep water we have a large value for h, i.e. as $h \Rightarrow \infty$ $\tanh(kh) \Rightarrow 1$ So we can see from Figure C.2 that the particles travel in a circle in deep water However in shallow water with decreasing h, i.e. as $h \Rightarrow 0$ $\tanh(kh) \Rightarrow 0$ The particles travel in an Ellipse which will become flatter, as shown in figure C.2 until they are eventually travelling in a flat line, as you see with waves on a beach.

C.4.4 Breaking Waves in Shallow Water

The velocity of a wave in shallow water can be approximated, for small h

$$\tanh(kh) \approx kh. \qquad (C.67)$$

Then

$$\omega = k\sqrt{gh}. \qquad (C.68)$$

Thus the velocity of the wave c, is approximated using equations (C.56) and (C.67) to

$$c = \sqrt{gh}. \qquad (C.69)$$

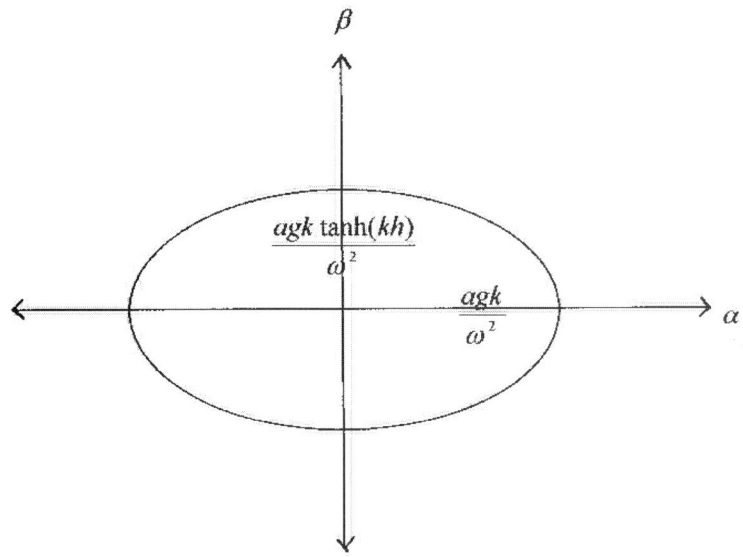

Figure C.2   Motion of the particles

Now the horizontal velocity component of $\underline{V}$, u in shallow water is approximated also, once again for small h

$$\cosh kh \approx 1 \qquad (C.70)$$

and so using (C.68), (C.70) our u from (C.59) becomes

$$u = \frac{ag \cos(kx - \omega t)}{\sqrt{gh}}. \qquad (C.71)$$

We can see that the velocity of the waves, in (C.69) decreases as the depth decreases but the velocity of the particle in (C.71) increases with decreasing depth. Therefore there is a point at which the velocity of the particles will begin to be faster than that of the wave.

When the velocities are equal (C.69 = C.71)

$$a \cos(kx - \omega t) = h. \qquad (C.72)$$

If this occurs at the top of a wave, the crest, i.e. $\cos(kx - \omega t) = 1$. Then the wave will break at, from (C.72), $h = a$ As u becomes greater then c the wave will *spill* over. The linear solution of η that we have obtained gives the crest

C. Project Example 3: Water Waves on a Sloping Beach

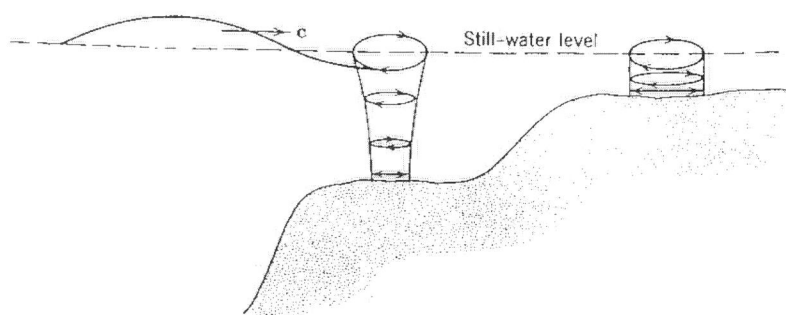

Figure C.3 The motion of a particle in shallow water. (Adapted from *Ocean Engineering Wave Mechanics* by M.E. McCormick, John Wiley and Sons Inc. Reprinted with permission.)

a sinusoidal shape, however as we can see from real waves on the beach it is evident that the crest takes on a slightly more pointed tip, as shown in Figure C.4 A non-linear solution for η would model this much better.

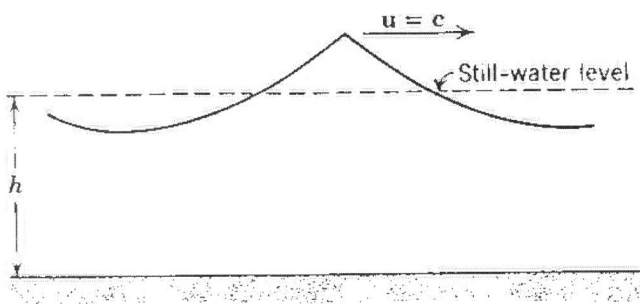

Figure C.4 The crest of a wave. (Adapted from *Ocean Engineering Wave Mechanics* by M.E. McCormick, John Wiley and Sons Inc. Reprinted with permission.)

C.5 [No Title]

C.5.1 Plane Waves

A plane wave is a 2 dimensional wave not now propagating along the x axis but in any direction, as shown in Figure C.5 We now have the coordinate system

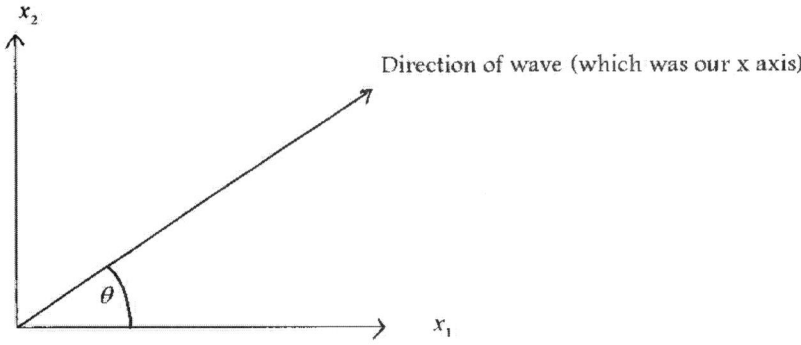

Figure C.5

(x_1, x_2, z) and the wave number k is represented as a vector k, where

$$k = (k\cos\theta, k\sin\theta) = (k_1, k_2). \qquad (C.73)$$

From this we need to find a new expression for kx in our Free Surface.

From Figure C.5 we can see that

$$x_1 = x\cos\theta \qquad (C.74)$$
$$x_2 = x\sin\theta \qquad (C.75)$$

which gives us x in terms of x_1 and x_2

$$x = x_1\cos\theta + x_2\sin\theta. \qquad (C.76)$$

Therefore we have

$$kx = x_1 k_1 + x_2 k_2 = k.x. \qquad (C.77)$$

The Free Surface of a plane wave is now expressed as

$$\eta = a\cos(k.x - \omega t) \qquad (C.78)$$

and Velocity Potential

$$\phi = A\cosh(kz + kh)\sin(k.x - \omega t) \qquad (C.79)$$

C. Project Example 3: Water Waves on a Sloping Beach

with corresponding vector wave velocity

$$c = \frac{\omega}{k^2} k. \tag{C.80}$$

C.5.2 Wave Rays

The equations obtained so far have been for an infinite train of uniform waves, but in reality this is not always the case. We are now going to allow small and slow changes in, wave amplitude, a, frequency, $\frac{\omega}{2\pi}$, wave number, k, and depth, h. The mean surface level previously assumed to be zero is also now a variable, say, b. So our depth is now

$$d = b + h. \tag{C.81}$$

So our Free Surface changes again to become

$$\eta = b + \cos \Phi \tag{C.82}$$

where the phase function

$$\Phi = k.x - \omega t \tag{C.83}$$

and

$$\phi = A \cosh(kz + kd) \sin(k.x - \omega t) \tag{C.84}$$
$$\tilde{\omega} = \sqrt{gk \tanh(kd)}. \tag{C.85}$$

We can find expressions for k_1, k_2 and ω from the phase function (C.83)

$$k_1 = \frac{\partial \Phi}{\partial x_1} \qquad k_2 = \frac{\partial \Phi}{\partial x_2} \qquad \omega = \frac{\partial \Phi}{\partial t}. \tag{C.86}$$

We can show that Φ holds for the slowly varying changes in the wave train talked about on the previous page. Using the Taylor Expansion near a fixed point (x_{10}, x_{20}, t_0)

$$\Phi(x_1, x_2, t) = \Phi(x_{10}, x_{20}, t_0) + (x_1 - x_{10})\frac{\partial \Phi}{\partial x_1}\bigg|_0 + (x_2 - x_{20})\frac{\partial \Phi}{\partial x_2}\bigg|_0 + (t - t_0)\frac{\partial \Phi}{\partial t}\bigg|_0 \tag{C.87}$$

(neglecting higher derivative terms). Using Equations (C.84) and (C.85) it can be shown that

$$\Phi = k_{10} x_1 + k_{20} x_2 - \omega_0 t. \tag{C.88}$$

Thus Φ hold locally.

It is assumed that the second derivatives of Φ are continuous, so Equations (C.84) give us

$$\frac{\partial k_1}{\partial x_2} = \frac{\partial k_2}{\partial x_1} \tag{C.89}$$

$$\frac{\partial k_1}{\partial t} + \frac{\partial \omega}{\partial x_1} = 0 \tag{C.90}$$

$$\frac{\partial k_2}{\partial t} + \frac{\partial \omega}{\partial x_2} = 0. \tag{C.91}$$

The depth is now a slowly varying function of x_1, x_2 and t. So ω can be represented as a function of (k_1, k_2, x_1, x_2, t)

$$\omega(k, d) = \Omega(k_1, k_2, x_1, x_2, t). \tag{C.92}$$

From the group velocity $c_g = \dfrac{d\omega}{dk}$ the vector group velocity could be assumed to be

$$c_g = \left(\frac{\partial \Omega}{\partial k_1}, \frac{\partial \Omega}{\partial k_2}\right) = (c_{g1}, c_{g2}). \tag{C.93}$$

Using Equations (C.90) and (C.92)

$$\frac{\partial \omega}{\partial x_1} = \frac{\partial \Omega}{\partial x_1} + \frac{\partial \Omega}{\partial k_1}\frac{\partial k_1}{\partial x_1} + \frac{\partial \Omega}{\partial k_2}\frac{\partial k_2}{\partial x_1} = -\frac{\partial k_1}{\partial t}. \tag{C.94}$$

Substituting Equation (C.89) and the vector group velocity (C.93), (C.94) can be written as

$$-\frac{\partial \Omega}{\partial x_1} = \frac{\partial k_1}{\partial t} + c_{g1}\frac{\partial k_1}{\partial x_1} + c_{g2}\frac{\partial k_1}{\partial x_2}. \tag{C.95}$$

Now the total derivative

$$\frac{Dk_1}{Dt} = \frac{\partial k_1}{\partial t} + (\underline{V}.\underline{\nabla})k_1 \tag{C.96}$$

$$\frac{Dk_1}{Dt} = \frac{\partial k_1}{\partial t} + u_1\frac{\partial k_1}{\partial x_1} + u_2\frac{\partial k_1}{\partial x_2}. \tag{C.97}$$

You can see from comparing (C.97) and (C.95) that the L.H.S. of (C.95) is also a total derivative but moving with the vector group velocity, c_g and not the particle velocity. We have

$$\frac{dx_1}{dt} = \frac{\partial \Omega}{\partial k_1} \tag{C.98}$$

$$\frac{dx_2}{dt} = \frac{\partial \Omega}{\partial k_2} \tag{C.99}$$

These define lines everywhere that are parallel to the vector group velocity and are called *Rays*

And on these *rays* from Equation (C.95)

$$\frac{dk_1}{dt} = \frac{\partial k_1}{\partial t} + \frac{dx_1}{dt}\frac{\partial k_1}{\partial x_1} + \frac{dx_2}{dt}\frac{\partial k_1}{\partial x_2} = -\frac{\partial \Omega}{\partial x_1}. \quad (C.100)$$

And from deducing $\frac{\partial \omega}{\partial x_2}$ and $\frac{\partial \omega}{\partial t}$ using the same method as in (C.94) we have on the rays that

$$\frac{dk_1}{dt} = -\frac{\partial \Omega}{\partial x_1} \quad (C.101)$$

$$\frac{dk_2}{dt} = -\frac{\partial \Omega}{\partial x_2} \quad (C.102)$$

$$\frac{d\omega}{dt} = \frac{\partial \Omega}{\partial t}. \quad (C.103)$$

From (C.92) the rays can also be written as

$$\frac{dx_1}{dt} = \frac{\partial \omega}{\partial k}\frac{k_1}{k} \quad (C.104)$$

$$\frac{dx_2}{dt} = \frac{\partial \omega}{\partial k}\frac{k_2}{k} \quad (C.105)$$

which is the form we will use in the next section.

These rays represent the direction of travel of the waves and the wave crests are perpendicular to this.

C.5.3 The Waves Approaching a Beach

Using the Ray equations established previously we can predict what will happen to the waves as the depth of water decreases, i.e. at a sloping beach. The figure below shows the sea in the x_1 less than zero region and the shore is the x_2 axis. To begin with we will look at the waves in the deep water, at $x_1 = -\infty$, before they are affected by the sloping beach. We shall also assume again that the variation in the mean surface level, b, is negligible and is zero. We shall assume there is no variation on depth, so Ω (C.92) is independent of x_1, x_2, and t which from (C.102) gives ω as a constant on the rays and so a constant everywhere. And in deep water using (C.65) we have

$$\omega^2 = gk. \quad (C.106)$$

The previous assumption also tells us that from (C.100) and (C.101), k_1 and k_2 are constant on the rays in deep water. Dividing the ray equations in (C.103)

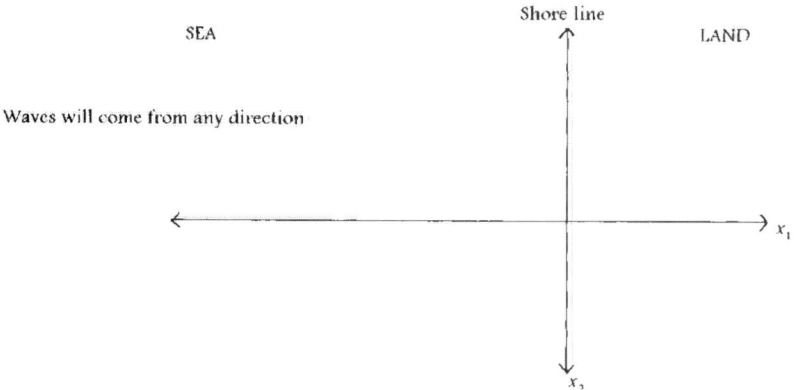

Figure C.6 Wave rays in deep water

we have
$$\frac{dx_2}{dx_1} = \frac{k_2}{k_1}. \tag{C.107}$$
And as k_1 and k_2 are constants we can integrate this
$$\int dx_2 = \int \frac{k_2}{k_1} dx_1 \tag{C.108}$$
$$x_2 = \frac{k_2}{k_1} x_1 + x_3 \tag{C.109}$$
where x_3 is the constant of integration. Equation (C.107) shows us that the waves in deep water travel in straight lines The wave crests are perpendicular to the rays so for the crests (C.105) can be written as
$$\frac{dx_2}{dx_1} = -\frac{k_1}{k_2} \tag{C.110}$$
which is integrated to give
$$x_2 = x_4 - \frac{k_1}{k_2} x_1 \tag{C.111}$$
where x_4 is the constant of integration.

These equations (C.107) and (C.109) tells us that in deep water the waves are not parallel to the shore (as the wave number is never zero), as shown in Figure C.7. But when standing on a beach the waves more often than not run up parallel to the beach, as in Figure C.9, so now we look at the wave rays when in shallow water.

C. Project Example 3: Water Waves on a Sloping Beach

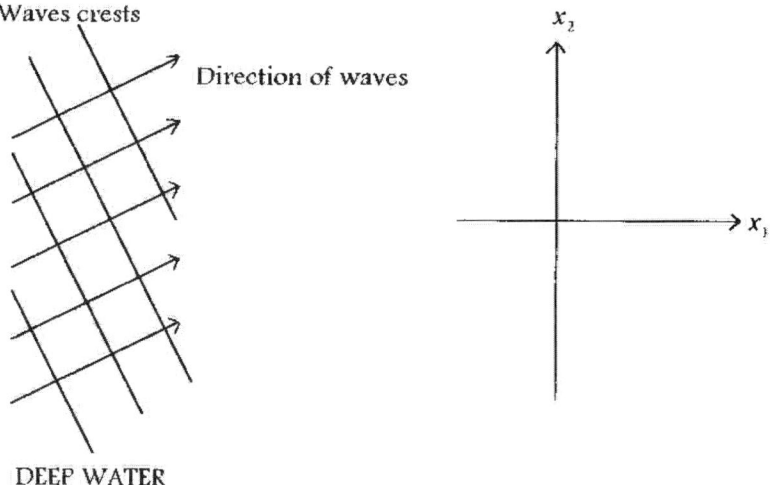

Figure C.7   Direction of waves in deep water

C.5.4 Wave rays in shallow water

We have established that ω is constant on these rays, so

$$\omega^2 = gk \tanh(kd) = \text{const.} \tag{C.112}$$

Thus as we approach the beach and the depth decreases, k must increase to keep ω^2 a constant, so if k increases the wave velocity

$$c = \frac{\omega}{k} \quad \text{will decrease, i.e. the waves will slow down}$$

and the wave length

$$\lambda = \frac{2\pi}{k} \quad \text{will also decrease, shortening the waves.}$$

We can see from this that as the wave nears the shore it will slow down and turn parallel to the shore, refraction of the waves.

We can write the depth of the water when it is shallow as a fraction of x_1 where the slope has a gradient of γ

$$d = -\gamma x_1. \tag{C.113}$$

From the shore, i.e. along the x_2 axis, the waves look the same, thus

$$\frac{\partial}{\partial x_2} = 0. \tag{C.114}$$

So from (C.101)
$$\frac{dk_2}{dt} = 0 \qquad (C.115)$$
$$\Rightarrow k_2 = \text{const} = m. \qquad (C.116)$$

Now that
$$k_2 = m \quad \Rightarrow k_1 = \sqrt{k^2 - m^2} \qquad (C.117)$$
and Equation (C.107) is now
$$\frac{dx_2}{dx_1} = \frac{m}{\sqrt{k^2 - m^2}} \qquad (C.118)$$

For small d
$$\tanh(kd) \sim kd. \qquad (C.119)$$
Using (C.119) and (C.113), (C.112) can be written as
$$\omega^2 = gk^2(-\gamma x_1). \qquad (C.120)$$
So k can be written as a function of x_1, and letting $x_1 = -|x_1|$
$$k^2 = \frac{\omega^2}{g\gamma|x_1|}. \qquad (C.121)$$

Thus the differential equation (C.118) is
$$\frac{dx_2}{dx_1} = \frac{m}{\sqrt{\frac{\omega}{g\gamma|x_1|} - m^2}}. \qquad (C.122)$$

And if $|x_1|$ is small enough
$$-\frac{dx_2}{d|x_1|} = m\sqrt{\frac{g\gamma|x_1|}{\omega^2}}. \qquad (C.123)$$

We can integrate (C.123) to give
$$x_2 = x_5 - \frac{2m\sqrt{g\gamma}|x_1|^{\frac{3}{2}}}{3\omega} \qquad (C.124)$$
(where x_5 is a constant of integration)

Once again we have for the crests
$$\frac{dx_2}{d|x_1|} = \frac{1}{m}\sqrt{\frac{\omega^2}{g\gamma|x_1|}}. \qquad (C.125)$$

Also integrated gives
$$x_2 = \frac{2\omega|x_1|^{\frac{1}{2}}}{m\sqrt{g\gamma}} - x_6 \qquad (C.126)$$

C. Project Example 3: Water Waves on a Sloping Beach

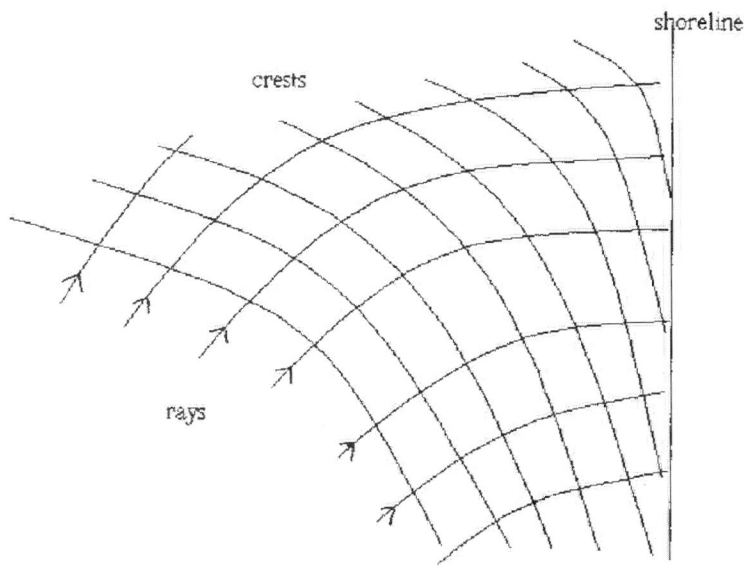

Figure C.8

(where x_6 is the constant of integration)

We now have from (C.124) and (C.126) the direction of travel of the waves and shape of the crests as $\omega, m, \gamma, g, x_5$ and x_6 are all constants. So we can draw the basic shape of these curves. Figure C.8 shows the basic form the wave rays and crests take approximated by (C.124) and (C.126) with varying constants of integration. This figure would obviously vary depending on the other constants. Comparing Figure C.8 to Figure C.9, real waves on a beach you can see that the solutions obtained are quite a good approximation to the actual waves.

C.6 [No Title]

C.6.1 Conclusion and Discussion

The solutions obtained, although only fairly simple approximations, I feel are good approximations to the direction of travel of waves on a beach, as can be seen by comparing Figures C.8 and C.9, even with the assumptions made.

I feel that the next thing to go on to here would be to look at a non-linear

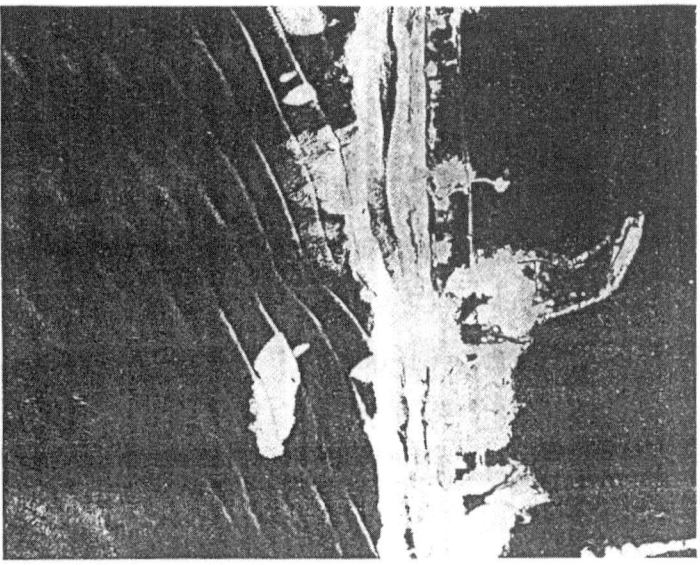

Figure C.9 Wave refraction at West Hampton Beach, Long Island. (Reproduced from *Shore Protection Manual*, Volume 1, Third edition, 1977, US Government Printing Office.)

Wave Theory, with which we could model the breaking waves on the beach to show the peaks the crests form. Stokes Theory for non-linear waves suggests letting the wave equations be represented as a series of small perturbation, the higher the order the more accurate an approximation is obtained.

There are many aspects of water waves that can be studied further and in more detail and I have only really touched the surface! Such things as the V shape pattern of waves generated by a moving object on the surface, whether the object is a streamlined speed boat or a lump of wood, or Kelvin's Theory of the wave pattern created by a moving ship. Another interesting wave is the solitary wave, a single 'bump' that travels uniformly.

References

1. Crapper, G.D. *Introduction to Water Waves* Ellis Horwood, Chichester, 1984.

2. McCormick, M.E. *Ocean Engineering Wave Mechanics* John Wiley and Sons Inc., New York, 1973.

3. LeBlond, P.H. and Mysak, L.A. *Waves in the Ocean* Elsevier, Amsterdam, 1978.

4. James, P.W. *MATH305 Course Notes*

5. Stoker, J.J. *Water Waves* Interscience, New York, 1957.

Epilogue

This student was one of the many that came to a lecturer and wanted to do "something about sailing"! The fundamental problem with this project is its lack of scope. Initially, the student and lecturer decided on trying to model the movement of floating bodies, hoping to treat the effects of wind on a sail in a rudimentary way. However, it soon became clear that this was not going to happen, and pragmatism restricted the project student to the study of water waves. Even given this, the project still failed to come up to expectations. Only elementary linear water wave theory was covered; there were no generalisations even though there was some relevant course material on hand. Something such as a description of the Gerstner wave, perturbation theory as used by Civil Engineers, even perhaps Cnoidal waves would have at least raised the level a little above the mundane, but no. This project ended up both limited in scope and short on relating water wave theory to lecture course material on inviscid flow or waves.

The style of the project itself was also rather immature. There were mistakes in the English (most of which remain), slips in the equations (most of which remain) and errors in the numbering of equations (all of which have been eliminated through the use of LaTeX). The layout of the project here is better than the original, again due to the use of LaTeX.

At the oral examination, things did not improve. The student was very quiet and gave a lack-luster performance. This seemed not to be due to nerves, more to a lack of preparation. It was almost as if the student realised that the project was irretrievably poor and had written it off. The most elementary questions were left unanswered.

There was enough here for a pass. The mathematics though elementary was (mostly) correct. The content though meagre was just acceptable. This student was unanimously but with some regret awarded a third class mark for this project. For the record, the student went on to gain an upper second on the strength of examination performance. Perhaps there really was a deliberate channelling of effort towards exams in this case.

Index

acceleration, Coriolis, 108
advice, 36
aerodynamic, 46
aerofoil, 47, 51
Airy's differential equation, 130
algebraic curves, 195
algorithm, 82, 88
angular momentum, 47
appeals, 19
applied mathematics, 2, 45, 62
argument, 73
assessment, 2, 11, 106
astronomy, 94
axioms, 37

background, non-historical, 60
Bessel's equation, 63
biography, 22, 31, 37
bipartite graph, 88
Black–Scholes equation, 96
Boolean algebra, 101
boomerang, 44, 45
boomerangs, 46, 51
branch points, 64

car density, 126
car velocity, 125
case studies, 10, 11, 24, 105
case study, 128
Cauchy integral formula, 133
characteristic curve, 127
characteristics, 123
characteristics, method of, 122
cheating, 1
circulation, 51

co-operation, 88
communication, 73
complex potential, 46
complex variable theory, 46, 128
computer algebra packages, 33
computer failure, 91
computer programs, 28
computing, 28, 65, 88
concepts, 129
confluent hypergeometric equation, 63
conics, 94
constructive feedback, 32
content, 66
content marks, 18
continuity equation, 124
continuous assessment, 1
contour integral solutions, 128
convergence, 143
convolution, 58
copying, 22, 35, 40
Coriolis acceleration, 48, 56, 106
Coriolis parameter, 56, 109
coursework, 5, 105
credits, 9, 21, 75
cross-ratio, 40

Davidon–Fletcher–Powell, 136, 139
debate, 73
Desargues' Theorem, 38
diffusion, 78, 80
diffusion equation, 79
Dirac-δ function, 90
discussion, 73
dividing into groups, 76
Duffing's equation, 116

dynamic meteorology, 53
dynamical systems, 116

earth, 107
eddy viscosity, 112, 113
Ekman equations, 113, 114, 147
Ekman flow, 114
Ekman layer, 55, 57
elliptic integral, 63
Euclid, 37
Euler, 141, 142
Euler constant, 145
Euler's equations, 48, 49
Eulerian view, 110
examinations, 2
Excel, 136
explanation, 84
explanation, not enough, 58
extension, 132
extensions, 115
external examiner, 23

feedback, 81
Fields Medal, vii
final report, 21, 81, 94
final year tutor, 23
financial mathematics, 96
fluid dynamics, 52
fluid mechanics, 46, 110
formal lectures, 10
formative, 12
FORTRAN, 28, 99
Fourier series, 89, 116
Fourier transform, 89, 90
Fourier transforms, 89
friction, 56, 111
frisbee, 45
frisbees, 51
Frobenius method, 59, 62

Galois Theory, 151
general manager, 90
geometric thinking, 43
geometry, 35, 38
geophysical fluid dynamics, 52
geostrophic balance, 55, 113
geostrophic flow, 57
geostrophic wind, 57
gradient, 138
grammar checkers, 33
graphs, 82
group project, 7, 11, 24, 74, 89, 95
group project presentation, 18
group velocity, 248

group work, 74
groups, 78
guidelines, 21

handout, 7, 74, 78, 89, 92, 93
handshaking lemma, 83
Hessian matrix, 138
Hill's equation, 120
historical research, 78
history of mathematics, 102, 141
homework, 11
homogeneous coordinates, 38
Hungarian algorithm, 84
hurricane, 52–54, 56
hypergeometric equation, 59
hypergeometric function, 59, 60, 62

independent assessor, 18
individual, 88
individual project, 3, 11, 32
industrial problem, 10
infinite series, 143
instant insanity, 84
integral operations, 129
interests, 31
interim report, 12, 13, 31, 91
inviscid flow, 50
iterative scheme, 138

job interviews, 35
Joukowski transform, 46, 50

Lagrangian mechanics, 92
Lagrangian view, 110
Laguerre's equation, 64
Laplace transform, 58
Laplace transforms, 89, 128
Laplace's linear equation, 128
Latex, 33
layout glitches, 95
Legendre's equation, 132, 133
levels of difficulty, 22
lift, 51
line at infinity, 38
linear waves, 120

management, 10
manual, 88
MAPLE, 33, 99, 134, 136
mathematical ability, 9, 75
mathematical accuracy, 22
mathematical biography, 22
mathematical content, 16
mathematical modelling, 65

mathematics education, 29
Mathieu's equation, 120, 148
mechanics, 44, 92
Microsoft Word, 33
mini-projects, 5, 82
model, 54
moderated, 24
moderating, 23
module leader, 9, 75, 80, 82, 91
modules, 105
moment of inertia, 47, 49

Navier Stokes' equation, 111
networks, 83
Newton–Raphson method, 135, 136, 139
Newtonian viscous fluid, 112
Nobel Prize, vii
non-linear first order differential
    equations, 116
non-linear optimisation, 141
non-linear oscillations, 116
normal modes, 97
numerical methods, 79
numerical solutions, 99
numerical techniques, 94

objectives, 36
ocean surface dynamics, 106
open book examination, 24
operational research, 135
opinion, 73
optimisation, 134, 137
oral examination, 40
oral presentation, 23
orbits, 93
overlap, 85

Pappus' theorem, 41, 42
parametric excitation, 120
partial differential equations, 55, 77, 134
partition, 77
peer assessment, 74, 81, 91
performance, 92
perturbation method, 118
perturbation solutions, 118
precession, 50
preference, 77
preliminary work, 5
presentation, 66, 80, 84, 88, 105
presentation marks, 18
presentation skills, 16
presentations, 91
presented material, 146
presenting, 74

pressure gradients, 112
Principle of Duality, 43
problem classes, 105
production problems, 95
Project Example, 151, 195, 241
project examples, 27, 66
project failure, 28
project list, 70
project report, 21
project supervisor, 18, 32
project team, 79
project titles, 35
projectiles, 45
projective geometry, 44
projectivity, 42
projects, 34
proof, 40–42, 142
proofs, 35
pure mathematics, 34, 36, 67

quality assessment, 76
quality assurance, 80
quality audit, 76

rate of change following the fluid, 111
real problem, 58
references, 33, 36
report writing, 31
research paper, 10
Reynolds stress, 111
Riemann P function, 59
rigid body, 46
rotating fluids, 55
rotating frame, 109

sandwich year, 3
Schläfli integrals, 133, 134
scholarly, 65
scope, 35, 45, 53, 60, 90
second assessor, 91
second marker, 23, 76
second order differential equation, 59
selecting a project, 29
shock, 127
shock wave, 122
skills, 65
special functions, 59
spell checkers, 33
stability analysis, 119
staff interests, 30
steepest descent, 138
strategies, 89
stress in a fluid, 111
student assignment, 139

student handbook, 31
subharmonic oscillations, 118
submission deadline, 32
supervisor, 12

technical level, 16
thermodynamics, 52, 55, 57
third assessor, 19, 23
tops and gyroscopes, 100
torque, 48
traditional lecture course, 105
traffic flow, 120
traffic jam, 124
traffic wave, 125
turbulence, 111

validation, 69
velocity potential, 247
verbal presentation, 13, 18, 94
vocational subjects, 3
vortex shedding, 51

water waves, 98, 249, 259
wave rays, 256
Whittaker's confluent hypergeometric function, 60
word processing, 33
writing skills, 28
writing style, 84
writing timetable, 32